THE ULTIMATE

HANDBOOK

A Complete Guide to Baking, Crafts, Gardening, Preserving Your Harvest, Raising Animals, and More

Abigail R. Gehring

Skyhorse Publishing

Skyhorse Publishing books may be purchased in bulk at special discounts for sales promotion, corporate gifts, fund-raising, or educational purposes. Special editions can also be created to specifications. For details, contact the Special Sales Department, Skyhorse Publishing, 307 West 36th Street, 11th Floor, New York, NY 10018 or info@skyhorsepublishing.com.

Visit our website at www.skyhorsepublishing.com.

10 9 8 7 6

Library of Congress Cataloging-in-Publication Data is available on file.
ISBN: 978-1-61608-710-4

Printed in China

Contents

v Introduction

Part 1

The Family Garden

2 Planning a Garden
9 Improving Your Soil
19 Planting Your Garden
23 Start Your Own Vegetable Garden
25 Growing Fruit Bushes and Trees
30 Growing and Threshing Grains
33 Container Gardening
38 Raised Beds
41 Pest and Disease Management
47 Harvesting Your Garden

Part 2

The Country Kitchen

52 Baking Bread
61 Maple Sugaring
64 Making Sausage
72 The Home Dairy
82 Canning
155 Drying and Freezing

Part 3

Country Crafts

166 Spring
175 Summer
184 Autumn
196 Winter

Part 4

The Barnyard

208 Chickens
212 Ducks
216 Turkeys
220 Rabbits
223 Beekeeping
228 Goats
233 Sheep
237 Llamas
241 Cows
245 Pigs
249 Butchering

254 Sources and Resources
259 Index

Introduction

More and more families are being drawn toward a lifestyle that is greener, cleaner, more genuine, and more aware. We want to know where our food is coming from, to the point of touching the dirt that it springs out of, if possible. We want our children (or nieces or nephews or godchildren) to understand that eggs come from chickens—not just from cardboard cartons on supermarket shelves. We love the idea of building things with our own hands, of picking our own berries, of making fresh bread and spreading it with homemade butter. We are, in short, longing for self-sufficiency.

"Self-sufficiency" as a term is somewhat misleading. "The good life" that most of us are seeking in our varied ways does not involve cutting off ties from those who surround us. Complete independence is not possible and, at least for most people, would not bring much satisfaction anyway. The early settlers banded together whenever they could, knowing their lives would be made easier and better by the community's support. In similar ways, we benefit from those who have ventured into back-to-basics living before us, and we would be wise to share ideas, tools, and experiences with those on similar paths around us now. But we do not need to be trapped by dependency on anyone or any group—or any idea, for that matter. We can be responsible for growing or raising at least a portion of what we consume; we can find ways to fix things rather than running to the store to buy replacements; we can teach our children ourselves, rather than leaving the burden entirely on public or private schools.

People and experience are the best teachers when it comes to learning things like how to plant a garden or milk a cow. But sometimes you don't have a neighbor to call on for advice and trial and error will result in more error than the trial is worth. That's where this book comes in. You'll find instructions and tips for everything from growing tomatoes to canning jams and jellies to constructing a chicken coop. Scattered throughout are fun projects for "The Junior Homesteader" and "Homeschooling Hints," which can be used to supplement your children's education, whether or not you choose to participate in a traditional schooling system. You'll also find plenty of photographs and illustrations to add clarity and interest to the written directions. Let these pages inspire and direct you as you discover what self-sufficiency means for you.

Part 1 The Family Garden

Planning a Garden	2
Improving Your Soil	9
Planting Your Garden	19
Start Your Own Vegetable Garden	23
Growing Fruit Bushes and Trees	25
Growing and Threshing Grains	30
Container Gardening	33
Raised Beds	38
Pest and Disease Management	41
Harvesting Your Garden	47

"It is utterly forbidden to be half-hearted about gardening. You have got to love your garden whether you like it or not."

—W.C. Sellar & R.J. Yeatman, Garden Rubbish, 1936

Planning a Garden

A Plant's Basic Needs

Before you start a garden, it's helpful to understand what plants need in order to thrive. Some plants, like dandelions, are tolerant of a wide variety of conditions, while others, such as orchids, have very specific requirements in order to grow successfully. Before spending time, effort, and money attempting to grow a new plant in a garden, learn about the conditions that particular plant needs in order to grow properly.

Environmental factors play a key role in the proper growth of plants. Some of the essential factors that influence this natural process are as follows:

1. Length of Day

The amount of time between sunrise and sunset is the most critical factor in regulating vegetative growth, blooming, flower development, and the initiation of dormancy. Plants utilize increasing day length as a cue to promote their growth in spring, while decreasing day length in fall prompts them to prepare for the impending cold weather. Many plants require specific day-length conditions in order to bloom and flower.

2. Light

Light is the energy source for all plants. Cloudy, rainy days or any shade cast by nearby plants and structures can significantly reduce the amount of light available to the plant. In addition, plants adapted to thrive in shady spaces cannot tolerate full

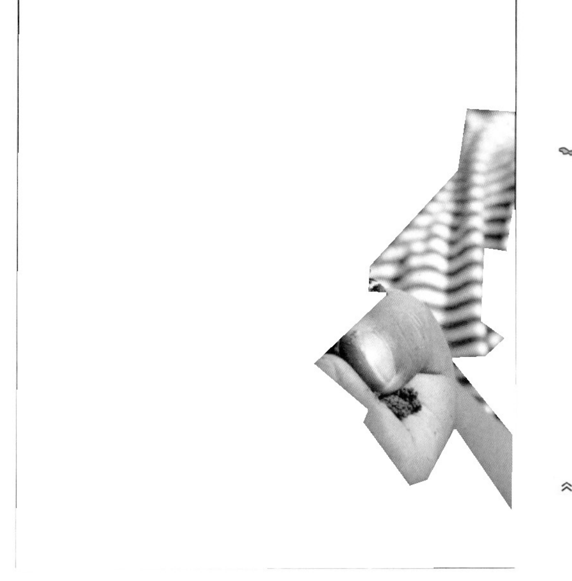

sunlight. In general, plants will only be able to survive where adequate sunlight reaches them at levels they are able to tolerate.

3. Temperature

Plants grow best within an optimal range of temperatures. This temperature range may vary drastically depending on the plant species. Some plants thrive in environments where the temperature range is quite wide; others can only survive within a very narrow temperature variance. Plants can only survive where temperatures allow them to carry on life-sustaining chemical reactions.

4. Cold

Plants differ by species in their ability to survive cold temperatures. Temperatures below 60°F injure some tropical plants. Conversely, arctic species can tolerate temperatures well below zero. The ability of a plant to withstand cold is a function of the degree of dormancy present in the plant and its general health. Exposure to wind, bright sunlight, or rapidly changing temperatures can also compromise a plant's tolerance to the cold.

5. Heat

A plant's ability to tolerate heat also varies widely from species to species. Many plants that evolved to grow in arid, tropical regions are

CREDIT: TIMOTHY W. LAWRENCE

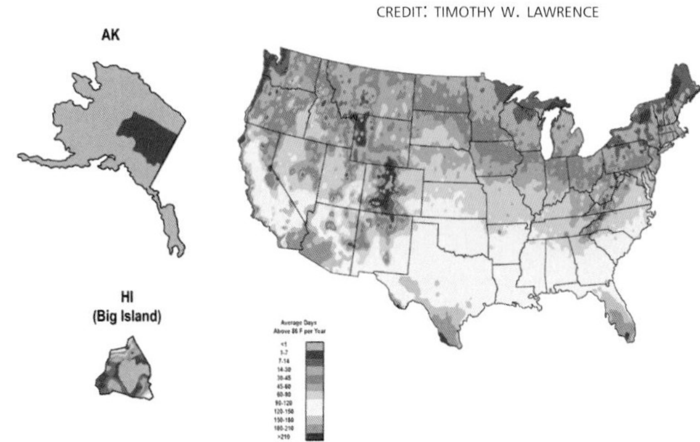

This map shows the average number of days each year that an area experiences temperatures over 86 degrees ("heat days"). Zone 1 has less than one heat day and Zone 12 has more than 210 heat days. Most plants begin to suffer when it gets any hotter than 86 degrees, though different plants have different levels of tolerance.

naturally very heat tolerant, while subarctic and alpine plants show very little tolerance for heat.

6. Water

Different types of plants have different water needs. Some plants can tolerate drought during the summer but need winter rains in order to flourish. Other plants need a consistent supply of moisture to grow well. Careful attention to a plant's need for supplemental water can help you to select plants that need a minimum of irrigation to perform well in your garden. If you have poorly drained, chronically wet soil, you can select lovely garden plants that naturally grow in bogs, marshlands, and other wet places.

7. Soil pH

A plant root's ability to take up certain nutrients depends on the pH—a measure of

A BASIC PLANT GLOSSARY

Annual—a plant that completes its life cycle in one year or season.

Arboretum—a landscaped space where trees, shrubs, and herbaceous plants are cultivated for scientific study or educational purposes, and to foster appreciation of plants.

Axil—the area between a leaf and the stem from which the leaf arises.

Bract—a leaflike structure that grows below a flower or cluster of flowers and is often colorful. Colored bracts attract pollinators, and are often mistaken for petals. Poinsettia and flowering dogwood are examples of plants with prominent bracts.

Cold hardy—capable of withstanding cold weather conditions.

Conifers—plants that predate true flowering plants in evolution; conifers lack true flowers and produce separate male and female strobili, or cones. Some conifers, such as yews, have fruits enclosed in a fleshy seed covering.

Cultivar—a cultivated variety of a plant selected for a feature that distinguishes it from the species from which it was selected.

Deciduous—having leaves that fall off or are shed seasonally to withstand adverse weather conditions, such as cold or drought.

Herbaceous—having little or no woody tissue. Most plants grown as perennials or annuals are herbaceous.

Hybrid—a plant, or group of plants, that results from the interbreeding of two distinct cultivars, varieties, species, or genera.

Inflorescence—a floral axis that contains many individual flowers in a specific arrangement; also known as a flower cluster.

Native plant—a plant that lives or grows naturally in a particular region without direct or indirect human intervention.

Panicle—a pyramidal, loosely branched flower cluster; a panicle is a type of inflorescence.

Perennial—persisting for several years, usually dying back to a perennial crown during the winter and initiating new growth each spring.

Shrub—a low-growing, woody plant, usually less than 15 feet tall, that often has multiple stems and may have a suckering growth habit (the tendency to sprout from the root system).

Taxonomy—the study of the general principles of scientific classification, especially the orderly classification of plants and animals according to their presumed natural relationships.

Tree—a woody perennial plant having a single, usually elongated main stem, or trunk, with few or no branches on its lower part.

Wildflower—a herbaceous plant that is native to a given area and is representative of unselected forms of its species.

Woody plant—a plant with persistent woody parts that do not die back in adverse conditions. Most woody plants are trees or shrubs.

the acidity or alkalinity—of your soil. Most plants grow best in soils that have a pH between 6.0 and 7.0. Most ericaceous plants, such as azaleas and blueberries, need acidic soils with pH below 6.0 to grow well. Lime can be used to raise the soil's pH, and materials containing sulfates, such as aluminum sulfate and iron sulfate, can be used to lower the pH. The solubility of many trace elements is controlled by pH, and plants can only use the soluble forms of these important micronutrients.

THE JUNIOR HOMESTEADER

You can structure your plants to double as playhouses for the kids. Here are a few possibilities:

Bean teepees are the best way to support pole bean plants, and they also make great hiding places for little gardeners. Drive five or six poles that are 7 to 8 feet tall into the ground in a circle with a 4-foot diameter. Bind the tops of the poles together with baling twine or a similar sturdy string. Plant your beans at the bottoms of the poles so they'll grow up and create a tent of vines.

Vine tunnels can be made out of poles and any trailing vines—gourds, cucumbers, or morning glories are a few options. Drive several 7- to 8-foot poles (bamboo works well) into the ground in two parallel lines so that they create a pathway. The poles should be at least 3 feet apart from each other. Then lash horizontal poles to the vertical ones at 2-, 4-, and 6-foot heights. You can also lash poles across the top of the tunnel to connect the two sides and create a roof. Plant your trailing vines at the bases of the poles and watch your tunnel fill in as the weeks go by.

Wigwams and huts are easily fashioned by planting your sunflowers or corn in a circular or square shape. To make a whole house, plant "walls" that are a few rows thick and create several "rooms," leaving gaps for doors.

Choosing a Site for Your Garden

Choosing the best spot for your garden is the first step toward growing the vegetables, fruits, and herbs that you want. You do not need a large space to get started—in fact, often it's wise to start small so that you don't get overwhelmed. A normal garden that is about 25 feet square will provide enough produce for a family of four, and with a little ingenuity (utilizing pots, hanging gardens, trellises, etc.) you can grow more than that in an even smaller space.

Five Factors to Consider When Choosing a Garden Site

1. Sunlight

Sunlight is crucial for the growth of vegetables and other plants. For your garden to grow, your plants will need at least six hours of direct sunlight per day. In order to make sure your garden receives an ample amount of sunlight, don't select a garden site that will be in the shade of trees, shrubs, houses, or other structures. Certain vegetables, such as broccoli and spinach, grow just fine in shadier spots, so if your garden does receive some shade, make sure to plant those types of vegetables in the shadier areas. However, on the whole, if your garden does not receive at least six hours of intense sunlight per day, it will not grow as efficiently or successfully.

2. Proximity

Think about convenience as you plot out your garden space. If your garden is closer

to your house and easy to reach, you will be more likely to tend it on a regular basis and to harvest the produce at its peak of ripeness. You'll find it a real boon to be able to run out to the garden in the middle of making dinner to pull up a head of lettuce or snip some fresh herbs.

3. Soil Quality

Your soil does not have to be perfect to grow a productive garden. However, it is best to have soil that is fertile, is full of organic materials that provide nutrients to the plant roots, and is easy to dig and till. Loose, well-drained soil is ideal. If there is a section of your yard where water does not easily drain after a good, soaking rain, this is not the spot for your garden; the excess water can easily drown your plants. Furthermore, soils that are of a clay or sandy consistency are not as effective in growing plants. To make these types of soils more nutrient-rich and fertile, add in organic materials (such as compost or manure).

4. Water Availability

Water is vital to keeping your garden green, healthy, and productive. A successful garden needs around 1 inch of water per week to thrive. Rain and irrigation systems are effective in maintaining this 1-inch-per-week quota. Situating your garden near a spigot or hose is ideal, allowing you to keep the soil moist and your plants happy.

5. Elevation

Your garden should not be located in an area where air cannot circulate or where frost quickly forms. Placing your garden in a low-lying area, such as at the base of a slope, should be avoided. Lower areas do not warm as quickly in the spring, and will easily collect frost in the spring and fall. Your garden should, if at all possible, be on a slightly higher elevation. This will help protect your plants from frost and you'll be able to start your garden growing earlier in the spring and harvest well into the fall.

Tools of the Trade

Gardening tools don't need to be high-tech, but having the right ones on hand will make your life much easier. You'll need a spade or digging fork to dig holes for seeds or seedlings (or, if the soil is loose enough, you can just use your hands). Use a trowel, rake, or hoe to smooth over the garden surface. A measuring stick is helpful when spacing your plants or seeds (if you don't have a measuring stick, you can use a precut string to measure). If you are planting seedlings or established plants, you may need stakes and string to tie them up so they don't fall over in inclement weather or when they start producing fruits or vegetables. Finally, if you are interested in installing an irrigation system for your garden, you will need to buy the appropriate materials for this purpose.

Companion Planting

Plants have natural substances built into their structures that repel or attract certain insects and can have an effect on the growth rate and even the flavor of the other plants around them. Thus, some plants aid each other's growth when planted in close proximity and others inhibit each other. Smart companion planting will help your garden remain healthy, beautiful, and in harmony, while deterring certain insect pests and other factors that could be potentially detrimental to your garden plants.

Below are charts that list various types of garden vegetables, herbs, and flowers and their respective companion and "enemy" plants.

If your garden is not close to your house, you may want to construct a small potting shed in which to keep your tools.

Shade-loving Plants

Most plants thrive on several hours of direct sunlight every day, but certain plants actually prefer the shade. When buying seedlings from your local nursery or planting your own seeds, read the accompanying label or packet before planting to make sure your plants will thrive in a shadier environment.

VEGETABLES

Type	Companion plant(s)	Avoid
Asparagus	Tomatoes, parsley, basil	Onions, garlic, potatoes
Beans	Eggplant	Tomatoes, onions kale
Beets	Mint	Runner beans
Broccoli	Onions, garlic, leeks	Tomatoes, peppers, mustard
Cabbage	Onions, garlic, leeks	Tomatoes, peppers, beans
Carrots	Leeks, beans	Radish
Celery	Daisies, snapdragons	Corn, aster flower
Corn	Legumes, squash, cucumbers	Tomatoes, celery
Cucumbers	Radish, beets, carrots	Tomatoes
Eggplant	Marigolds, mint	Runner beans
Leeks	Carrots	Legumes
Lettuce	Radish, carrots	Celery, cabbage, parsley
Melon	Pumpkin, squash	None
Onions	Carrots	Peas, beans
Peas	Beans, corn	Onions, garlic
Peppers	Tomatoes	Beans, cabbage, kale
Potatoes	Horseradish	Tomatoes, cucumbers
Tomatoes	Carrots, celery, parsley	Corn, peas, potatoes, kale

HERBS

Type	Companion Plant(s)	Avoid
Basil	Chamomile, anise	Sage
Chamomile	Basil, cabbage	Other herbs (it will become oily)
Chives	Carrots	Peas, beans
Cilantro	Beans, peas	None
Dill	Cabbage, cucumbers	Tomatoes, carrots
Fennel	Dill	Everything else
Garlic	Cucumbers, peas, lettuce	None
Oregano	Basil, peppers	None
Peppermint	Broccoli, cabbage	None
Rosemary	Sage, beans, carrots	None
Sage	Rosemary, beans	None
Summer savory	Onions, green beans	None

Bleeding heart plants thrive in shady areas.

FLOWERS

Types	Companion Plant(s)	Avoid
Geraniums	Roses, tomatoes	None
Marigolds	Tomatoes, peppers, most plants	None
Petunia	Squash, asparagus	None
Sunflower	Corn, tomatoes	None
Tansy	Roses, cucumbers, squash	None

Flowering plants that do well in partial and full shade include:

Bee balm
Bellflower
Bleeding heart
Cardinal flower
Coleus
Columbine
Daylily
Dichondra
Ferns
Forget-me-not
Globe daisy
Golden bleeding heart
Impatiens
Leopardbane
Lily of the valley
Meadow rue
Pansy
Periwinkle
Persian violet
Primrose
Rue anemone
Snapdragon
Sweet alyssum
Thyme

Vegetable plants that can grow in partial shade include:

Arugula
Beans
Beets
Broccoli
Brussels sprouts
Cauliflower
Endive
Kale
Leaf lettuce
Peas
Radish
Spinach
Swiss chard

When planting gardens against your home, choose shade-loving plants.

JUNIOR HOMESTEADER TIP

Kids as young as toddlers will enjoy being involved in a family garden. Encourage very young children to explore by touching and smelling dirt, leaves, and flowers. Just be careful they don't taste anything non-edible. If space allows, assign a small plot for older children to plant and tend all on their own. An added bonus is that kids are more likely to eat vegetables they've grown themselves!

Improving Your Soil

Nutrient-rich, fertile soil is essential for growing the best and healthiest plants—plants that will supply you with quality fruits, vegetables, and flowers. Sometimes soil loses its fertility (or has minimum fertility based on the region in which you live), so measures must be taken in order to improve your soil and, subsequently, your garden.

Soil Quality Indicators

Soil quality is an assessment of how well soil performs all of its functions now and how those functions are being preserved for future use. The quality of soil cannot just be determined by measuring row or garden yield, water quality, or any other single outcome, nor can it be measured directly. Thus, it is important to look at specific indicators to better understand the properties of soil. Plants can provide us with clues about how well the soil is functioning—whether a plant is growing and producing quality fruits and vegetables, or failing to yield such things, is a good indicator of the quality of the soil it's growing in.

In short, indicators are measurable properties of soil or plants that provide clues about how well the soil can function. Indicators can be physical, chemical, and biological properties, processes, or characteristics of soils. They can also be visual features of plants.

Useful indicators of soil quality:

- are easy to measure
- measure changes in soil functions
- encompass chemical, biological, and physical properties
- are accessible to many users
- are sensitive to variations in climate and management

Indicators can be assessed by qualitative or quantitative techniques, such as soil tests. After measurements are collected, they can be evaluated by looking for patterns and comparing results to measurements taken at a different time.

Examples of soil quality indicators:

1. Soil Organic Matter—promotes soil fertility, structure, stability, and nutrient retention and helps combat soil erosion.
2. Physical Indicators—these include soil structure, depth, infiltration and bulk density, and water hold capacity.

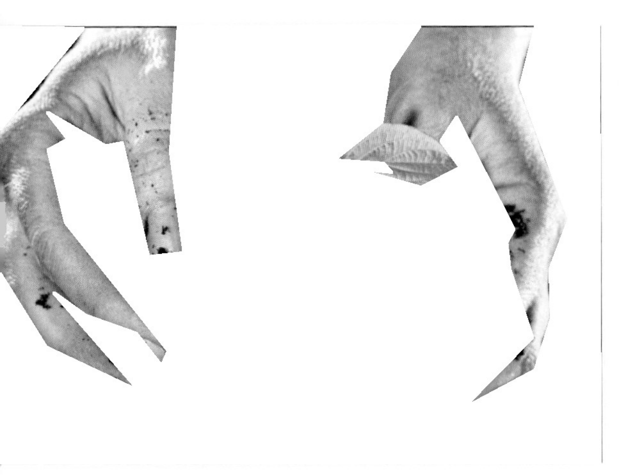

Quality soil will retain and transport water and nutrients effectively; it will provide habitat for microbes; it will promote compaction and water movement; and, it will be porous and easy to work with.

3. Chemical Indicators—these include pH, electrical conductivity, and extractable nutrients. Quality soil will be at its threshold for plant, microbial, biological, and chemical activity; it will also have plant nutrients that are readily available.
4. Biological Indicators—these include microbial biomass, mineralizable nitrogen, and soil respiration. Quality soil is a good repository for nitrogen and other basic nutrients for prosperous plant growth; it has a high soil productivity and nitrogen supply; and there is a good amount of microbial activity.

Soil and Plant Nutrients

Nutrient Management

There are twenty nutrients that all plants require. Six of the most important nutrients, called macronutrients, are: calcium, magnesium, nitrogen, phosphorus, potassium, and sulfur. Of these, nitrogen, phosphorus, and potassium are essential to healthy plant growth and are required in relatively large amounts. Nitrogen is associated with lush vegetative growth, phosphorus is required for flowering and fruiting, and potassium is necessary for durability and disease resistance. Calcium, sulfur, and magnesium are also required in comparatively large quantities and aid in the overall health of plants.

The other nutrients, referred to as micronutrients, are required in very small amounts. These include such elements as copper, zinc, iron, and boron. While both macro- and micronutrients are required for good plant growth, overapplication of these nutrients can be as detrimental to the plant as a deficiency of them. Overapplication of plant nutrients may not only impair plant growth, but may also contaminate groundwater by penetrating through the soil or may pollute surface waters.

Soil Testing

Testing your soil for nutrients and pH is important in order to provide your plants with the proper balance of nutrients (while avoiding overapplication). If you are establishing a new lawn or garden, a soil test is strongly recommended. The cost of soil testing is minor in comparison to the cost of plant materials and labor. Correcting a problem before planting is much simpler and cheaper than afterwards.

Once your garden is established, continue to take periodic soil samples. While many people routinely lime their soil, this can raise the pH of the soil too high. Likewise, since many fertilizers tend to lower the soil's pH, it may drop below desirable levels after several years, depending on fertilization and other soil factors, so occasional testing is strongly encouraged.

Home tests for pH, nitrogen, phosphorus, and potassium are available from most garden centers. While these may give you a general idea of the nutrients in your soil, they are not as reliable as tests performed by the Cooperative Extension Service at land grant universities. University and other commercial testing services will provide more detail, and you can request special tests for micronutrients if you suspect a problem. In addition to the analysis of nutrients in your soil, these services often provide recommendations for the application of nutrients or how best to adjust the pH of your soil.

The test for soil pH is very simple. Basically, pH is a measure of how acidic or alkaline your soil is. A pH of 7 is considered neutral. Below 7 is acidic and above 7 is alkaline. Since pH greatly influences plant nutrients, adjusting the pH will often correct a nutrient problem. At a high pH, several of the micronutrients become less available for plant uptake. Iron deficiency is a common problem, even at a neutral pH, for such plants as rhododendrons and blueberries. At a very low soil pH, other micronutrients may be too available to the plant, resulting in toxicity.

Phosphorus and potassium are tested regularly by commercial testing labs. While there are soil tests for nitrogen, these may be less reliable. Nitrogen is present in the soil in several forms that can change rapidly. Therefore, a precise analysis of nitrogen is more difficult to obtain. Most university soil test labs do not routinely test for nitrogen. Home testing kits often contain a test for nitrogen that may give you a general—though not necessarily completely accurate—idea of the presence of nitrogen in your garden soil.

Organic matter is often part of a soil test. Organic matter has a large influence on soil structure and so is highly desirable for your garden soil. Good soil structure improves aeration, water movement, and retention. This encourages increased microbial activity and root growth, both of which influence the availability of nutrients for plant growth. Soils high in organic matter tend to have a greater supply of plant nutrients compared to many soils low in organic matter. Organic matter tends to bind up some soil pesticides, reducing their effectiveness, and so this should be taken into consideration if you are planning to apply pesticides in your garden.

Tests for micronutrients are usually not performed unless there is reason to suspect a problem. Certain plants have greater requirements for specific micronutrients and may show deficiency symptoms if those nutrients are not readily available.

STEPS FOR TAKING A SOIL TEST

- If you intend to send your sample to the land grant university in your state, contact the local Cooperative Extension Service for information and sample bags. If you intend to send your sample to a private testing lab, contact them for specific details about submitting a sample.
- Follow the directions carefully for submitting the sample. The following are general guidelines for taking a soil sample:
  1. Sample when the soil is moist but not wet.
  2. Obtain a clean pail or similar container.
  3. Clear away the surface litter or grass.
  4. With a spade or soil auger, dig a small amount of soil to a depth of 6 inches.
  5. Place the soil in the clean pail.
  6. Repeat steps 3 through 5 until the required number of samples has been collected.
  7. Mix the samples together thoroughly.
  8. From the mixture, take the sample that will be sent for analysis.
  9. Send immediately. Do not dry before sending.
- If you are using a home soil testing kit, follow the above steps for taking your sample. Follow the directions in the test kit carefully so you receive the most accurate reading possible.

Enriching Your Soil

Organic and Commercial Fertilizers and Returning Nutrients to Your Soil

Once you have the results of the soil test, you can add nutrients or soil amendments as needed to alter the pH. If you need to raise the soil's pH, use lime. Lime is most effective when it is mixed into the soil; therefore, it is best to apply before planting (if you apply lime in the fall, it has a better chance of correcting any soil acidity problems for the next growing season). For large areas, rototilling is most effective. For small areas or around plants, working the lime into the soil with a spade or cultivator is preferable. When working around plants, be careful not to dig too deeply or roughly so that you damage plant roots. Depending on the form of lime and the soil conditions, the change in pH may be gradual. It may take several months before a significant change is noted. Soils high in organic matter and clay tend to take larger amounts of lime to change the pH than do sandy soils.

If you need to lower the pH significantly, especially for plants such as rhododendrons, you can use aluminum sulfate. In all cases, follow the soil test or manufacturer's recommended rates of application. Again, mixing well into the soil is recommended.

There are numerous choices for providing nitrogen, phosphorus, and potassium, the nutrients your plants need to thrive. Nitrogen (N) is needed for healthy, green growth and regulation of other nutrients. Phosphorus (P) helps roots and seeds properly develop and resist disease. Potassium (K) is also important in root development and disease resistance. If your soil is of adequate fertility, applying compost may be the best method of introducing additional nutrients. While compost is relatively low in nutrients compared to commercial fertilizers, it is especially beneficial in improving the condition of the soil and is nontoxic. By keeping the soil loose, compost allows plant roots to grow well throughout the soil, helping them to extract nutrients from a larger area. A loose soil enriched with compost is also an excellent habitat for earthworms and other beneficial soil microorganisms that are essential for releasing nutrients for plant use. The nutrients from compost are also released slowly, so there is no concern about "burning" the plant with an overapplication of synthetic fertilizer.

Manure is also an excellent source of plant nutrients and is an organic matter. Manure should be composted before applying, as fresh manure may be too strong and can injure plants. Be careful when composting manure. If left in the open, exposed to rain, nutrients may leach out of the manure and the runoff can contaminate nearby waterways. Make sure the manure is stored in a location away from wells and any waterways and that any runoff is confined or slowly released into a vegetated area. Improperly applied manure also can be a source of pollution. If you are not composting your own manure, you can purchase some at your local garden store. For best results, work composted manure into the soil around the plants or in your garden before planting.

If preparing a bed before planting, compost and manure may be worked into the soil to a depth of 8 to 12 inches. If adding to existing plants, work carefully around the plants so as not to harm the existing roots.

Green manures are another source of organic matter and plant nutrients. Green manures are crops that are grown and then tilled into the soil. As they break down, nitrogen and other plant nutrients become available. These manures may also provide additional benefits of reducing soil erosion. Green manures, such as rye and oats, are often planted in the fall after the crops have been harvested. In the spring, these are tilled under before planting.

With all organic sources of nitrogen, whether compost or manure, the nitrogen must be changed to an inorganic form before the plants can use it. Therefore, it is important to have well-drained, aerated soils that provide the favorable habitat for the soil microorganisms responsible for these conversions.

There are also numerous sources of commercial fertilizers that supply nitrogen, phosphorus, and potassium, though it is preferable to use organic fertilizers, such as compost and manures. However, if you choose to use a commercial fertilizer, it is important to know how to read the amount of nutrients contained in each bag. The first number on the fertilizer analysis is the percentage of nitrogen; the second number is phosphorus; and the third number is the potassium content. A fertilizer that has a 10-20-10 analysis contains twice as much of each of the nutrients as a 5-10-5. How much of each nutrient you need depends on your soil test results and the plants you are fertilizing.

As mentioned before, nitrogen stimulates vegetative growth while phosphorus stimulates flowering. Too much nitrogen can inhibit flowering and fruit production. For many flowers and vegetables, a fertilizer higher in phosphorus than nitrogen is preferred, such as a 5-10-5. For lawns, nitrogen is usually required in greater amounts, so a fertilizer with a greater amount of nitrogen is more beneficial.

Fertilizer Application

Commercial fertilizers are normally applied as a dry, granular material or mixed with water

Soil Test Reading	What to Do
High pH	Your soil is alkaline. To lower pH, add elemental sulfur, gypsum, or cottonseed meal. Sulfur can take several months to lower your soil's pH, as it must first convert to sulfuric acid with the help of the soil's bacteria.
Low pH	Your soil is too acidic. Add lime or wood ashes.
Low nitrogen	Add manure, horn or hoof meal, cottonseed meal, fish meal, or dried blood.
High nitrogen	Your soil may be overfertilized. Water the soil frequently and don't add any fertilizer.
Low phosphorus	Add cottonseed meal, bone meal, fish meal, rock phosphate, dried blood, or wood ashes.
High phosphorus	Your soil may be overfertilized. Avoid adding phosphorus-rich materials and grow lots of plants to use up the excess.
Low potassium	Add potash, wood ashes, manure, dried seaweed, fish meal, or cottonseed meal.
High potassium	Continue to fertilize with nitrogen and phosphorus-rich soil additions, but avoid potassium-rich fertilizers for at least two years.
Poor drainage or too much drainage	If your soil is a heavy, clay-like consistency, it won't drain well. If it's too sandy, it won't absorb nutrients as it should. Mix in peat moss or compost to achieve a better texture.

HOW TO PROPERLY APPLY FERTILIZER TO YOUR GARDEN

Apply fertilizer when the soil is moist, and then water lightly. This will help the fertilizer move into the root zone where its nutrients are available to the plants, rather than staying on top of the soil where it can be blown or washed away.

Watch the weather. Avoid applying fertilizer immediately before a heavy rain system is predicted to arrive. Too much rain (or sprinkler water) will take the nutrients away from the lawn's root zone and could move the fertilizer into another water system, contaminating it.

Use the minimum amount of fertilizer necessary and apply it in small, frequent applications. An application of two pounds of fertilizer, five times per year, is better than five pounds of fertilizer twice a year.

If you are spreading the fertilizer by hand in your garden, wear gardening gloves and be sure not to damage the plant or roots around which you are fertilizing.

and poured onto the garden. If using granular materials, avoid spilling on sidewalks and driveways because these materials are water soluble and can cause pollution problems if rinsed into storm sewers. Granular fertilizers are a type of salt, and if applied too heavily, they have the capability of burning the plants. If using a liquid fertilizer, apply directly to or around the base of each plant and try to contain it within the garden only.

In order to decrease the potential for pollution and to gain the greatest benefits from fertilizer, whether it's a commercial variety, compost, or other organic materials, apply it when the plants have the greatest need for the nutrients. Plants that are not actively growing do not have a high requirement for nutrients; thus, nutrients applied to dormant plants, or plants growing slowly due to cool temperatures, are more likely to be wasted. While light applications of nitrogen may be recommended for lawns in the fall, generally, nitrogen fertilizers should not be applied to most plants in the fall in regions of the country that experience cold winters. Since nitrogen encourages vegetative growth, if it is applied in the fall it may reduce the plant's ability to harden properly for winter.

In some gardens, you can reduce fertilizer use by applying it around the individual plants rather than broadcasting it across the entire garden. Much of the phosphorus in fertilizer becomes unavailable to the plants once spread on the soil. For better plant uptake, apply the fertilizer in a band near the plant. Do not apply directly to the plant or in contact with the roots, as it may burn and damage the plant and its root system.

Rules of Thumb for Proper Fertilizer Use

It is easiest to apply fertilizer before or at the time of planting. Fertilizers can either be spread over a large area or confined to garden rows, depending on the condition of your soil and the types of plants you will be growing. After spreading, till the fertilizer into the soil about 3 to 4 inches deep. Only spread about one half of the fertilizer this way and then dispatch the rest 3 inches to the sides of each row and also a little below each seed or established plant. This method, minus the spreader, is used when applying fertilizer to specific rows or plants by hand.

Composting in Your Backyard

Composting is nature's own way of recycling yard and household wastes by converting them into valuable fertilizer, soil organic matter, and a source of plant nutrients. The result of this controlled decomposition of organic matter—a dark, crumbly, earthy-smelling material—works wonders on all kinds of soil by providing vital nutrients, and contributing to good aeration and moisture-holding capacity, to help plants grow and look better.

Composting can be as simple or as involved as you would like, depending on how much yard waste you have, how fast you want results, and the effort you are willing to invest. Since all organic matter eventually decomposes, composting speeds up the process by providing an ideal environment for bacteria and other decomposing microorganisms. The composting season coincides with the growing season, when conditions are favorable for plant growth—those same conditions work well for biological activity in the compost pile. However, since compost generates heat, the process may continue later into the fall or winter. The final product—called humus or compost—looks and feels like fertile garden soil.

Compost Preparation

While a multitude of organisms, fungi, and bacteria are involved in the overall process, there are four basic ingredients for composting: nitrogen, carbon, water, and air.

A wide range of materials may be composted because anything that was once alive will naturally decompose. The starting materials for composting, commonly referred to as feed stocks, include leaves, grass clippings, straw, vegetable and fruit scraps, coffee grounds, livestock manure, sawdust, and shredded paper. However, some materials that always should be avoided include diseased plants, dead animals, noxious weeds, meat scraps that may attract animals, and dog or cat manure, which can carry disease. Since adding kitchen wastes to compost may attract flies and insects, make a hole in the center of your pile and bury the waste.

For best results, you will want an even ratio of green, or wet, material, which is high in nitrogen, and brown, or dry, material, which is high in carbon. Simply layer or mix landscape trimmings and grass clippings, for example, with dried leaves and twigs in a pile or enclosure. If there is not a good supply of nitrogen-rich material, a handful of general lawn fertilizer or barnyard manure will help even out the ratio.

Though rain provides the moisture, you may need to water the pile in dry weather or cover it in extremely wet weather. The microorganisms in the compost pile function best when the materials are as damp as a wrung-out sponge—not saturated with water. A moisture content of 40 to 60 percent is preferable. To test for adequate moisture, reach into your compost pile, grab a handful of material, and squeeze it. If a few drops of water come out, it probably has enough moisture. If it doesn't, add water by putting a hose into the pile so that you aren't just wetting the top, or, better yet, water the pile as you turn it.

Air is the only part that cannot be added in excess. For proper aeration, you'll need to punch holes in the pile so it has many air passages. The air in the pile is usually used up faster than the moisture, and extremes of sun or rain can adversely affect this balance, so the materials must be turned or mixed up often with a pitchfork, rake, or other garden tool to add air that will sustain high temperatures, control odor, and yield faster decomposition.

Over time, you'll see that the microorganisms, which are small forms of plant and animal life, will break down the organic material. Bacteria are the first to break down plant tis-

sue and are the most numerous and effective compost-makers in your compost pile. Fungi and protozoans soon join the bacteria and, somewhat later in the cycle, centipedes, millipedes, beetles, sow bugs, nematodes, worms, and numerous others complete the composting process. With the right ingredients and favorable weather conditions, you can have a finished compost pile in a few weeks.

How to Make Your Own Backyard Composting Heap

1. Choose a level, well-drained site, preferably near your garden.
2. Decide whether you will be using a bin after checking on any local or state regulations for composting in urban areas, as some communities require rodent-proof bins. There are numerous styles of compost bins available, depending on your needs, ranging from a moveable bin formed by wire mesh to a more substantial wooden structure consisting of several compartments. You can easily make your own bin using chicken wire or scrap wood. While a bin will help contain the pile, it is not absolutely necessary, as you can build your pile directly on the ground. To help with aeration, you may want to place some woody material on the ground where you will build your pile.
3. Ensure that your pile will have a minimum dimension of 3 feet all around, but is no taller than 5 feet, as not enough air will reach the microorganisms at the center if it is too tall. If you don't have this amount at one time, simply stockpile your materials until a sufficient quantity is available for proper mixing. When composting is completed, the total volume of the original materials is usually reduced by 30 to 50 percent.
4. Build your pile by using either alternating equal layers of high-carbon and high-nitrogen material or by mixing equal parts of both together and then heaping it into a pile. If you choose to alternate layers, make each layer 2 to 4 inches thick. Some composters find that mixing the two together is more effective than layering. Adding a few shovels of soil will also help get the pile off to a good start because soil adds commonly found, decomposing organisms to your compost.
5. Keep the pile moist but not wet. Soggy piles encourage the growth of organisms

COMMON COMPOSTING MATERIALS

Cardboard
Coffee grounds
Corn cobs
Corn stalks
Food scraps
Grass clippings
Hedge trimmings
Livestock manure
Newspapers
Old potting soil
Pine needles
Plant stalks
Sawdust
Seaweed
Shredded paper
Straw
Tea bags
Telephone books
Tree leaves and twigs
Vegetable scraps
Weeds without seed heads
Wood chips
Woody brush

Avoid using:

Bread and grains
Cooking oil
Dairy products
Dead animals
Diseased plant material
Dog or cat manure
Grease or oily foods
Meat or fish scraps
Noxious or invasive weeds
Weeds with seed heads

that can live without oxygen and cause unpleasant odors.

6. Punch holes in the sides of the pile for aeration. The pile will heat up and then begin to cool. The most efficient decomposing bacteria thrive in temperatures between 110 and 160 degrees Fahrenheit. You can track this with a compost thermometer, or you can simply reach into the pile to determine if it is uncomfortably hot to the touch. At these temperatures, the pile kills most weed seeds and plant diseases. However, studies have shown that compost produced at these temperatures has less ability to suppress diseases in the soil, since these temperatures may kill some of the beneficial bacteria necessary to suppress disease.
7. Check your bin regularly during the composting season to assure optimum moisture and aeration are present in the material being composted.
8. Move materials from the center to the outside of the pile and vice versa. Turn every day or two and you should get compost in less than four weeks. Turning every other week will make compost in one to three months. Finished compost will smell sweet and be cool and crumbly to the touch.

Other Types of Composting

Cold or Slow Composting

Cold composting allows you to pile just organic material on the ground or in a bin. This method requires no maintenance, but it will take several months to a year or more for the pile to decompose, though the process is faster in warmer climates than in cooler areas. Cold composting works well if you are short on time needed to tend to the compost pile at least every other day, have little yard waste, and are not in a hurry to use the compost.

For this method, add yard waste as it accumulates. To speed up the process, shred or chop the materials by running over small piles of trimmings with your lawn mower, because the more surface area the microorganisms have to feed on, the faster the materials will break down.

Cold composting has been shown to be better at suppressing soil-borne diseases than hot composting and also leaves more non-decomposed bits of material, which can be screened out if desired. However, because of the low temperatures achieved during decomposition, weed seeds and disease-causing organisms may not be destroyed.

Vermicomposting

Vermicomposting uses worms to compost. This takes up very little space and can be done year-round in a basement or garage. It is an excellent way to dispose of kitchen wastes.

Here's how to make your own vermicomposting pile:

1. Obtain a plastic storage bin. One bin measuring 1 foot by 2 feet by 3½ feet will be enough to meet the needs of a family of six.
2. Drill eight to ten holes about ¼ inch in diameter in the bottom of the bin for drainage.
3. Line the bottom of the bin with a fine nylon mesh to keep the worms from escaping.
4. Put a tray underneath to catch the drainage.
5. Rip shredded newspaper into pieces to use as bedding and pour water over the strips until they are thoroughly moist. Place these shredded bits on one side of your bin. Do not let them dry out.
6. Add worms to your bin. It's best to have about two pounds of worms (roughly

THE JUNIOR HOMESTEADER

Let the kids be in charge of feeding the worms in your compost. They'll be fascinated by the squirmy critters!

Any large bucket can be turned into a compost barrel. You can cut out a piece of the barrel for easy access to the compost, as shown here, or simply access the compost through the lid. Drilling holes in the sides and lid of the bucket will increase air circulation and speed up the process. Leave your bucket in the sun and shake it, roll it, or stir the contents regularly.

2,000 worms) per one pound of food waste. You may want to start with less food waste and increase the amount as your worm population grows. Redworms are recommended for best composting, but other species can be used. Redworms are the common, small worms found in most gardens and lawns. You can collect them from under a pile of mulch or order them from a garden catalog.

7. Provide worms with food wastes such as vegetable peelings. Do not add fat or meat products. Limit their feed, as too much at once may cause the material to rot.
8. Keep the bin in a dark location away from extreme temperatures.
9. Wait about three months and you'll see that the worms have changed the bedding and food wastes into compost. At this time, open your bin in a bright light and the worms will burrow into the bedding. Add fresh bedding and more food to the other side of the bin. The worms should migrate to the new food supply.
10. Scoop out the finished compost and apply to your plants or save to use in the spring.

Common Problems

Composting is not an exact science. Experience will tell you what works best for you. If you notice that nothing is happening, you may need to add more nitrogen, water, or air; chip or grind the materials; or adjust the size of the pile.

If the pile is too hot, you probably have too much nitrogen and need to add additional carbon materials to reduce the heating.

A bad smell may indicate not enough air or too much moisture. Simply turn the pile or add dry materials to the wet pile to get rid of the odor.

Uses for Compost

Compost contains nutrients, but it is not a substitute for fertilizers. Compost holds nutrients in the soil until plants can use them, loosens and aerates clay soils, and retains water in sandy soils.

To use as a soil amendment, mix 2 to 5 inches of compost into vegetable and flower gardens each year before planting. In a potting mixture, add one part compost to two parts commercial potting soil, or make your own mixture by using equal parts of compost and sand or perlite.

As a mulch, spread an inch or two of compost around annual flowers and vegetables, and up to 6 inches around trees and shrubs. Studies have shown that compost used as mulch, or mixed with the top 1-inch layer of soil, can help prevent some plant diseases, including some of those that cause damping of seedlings.

As a top dressing, mix finely sifted compost with sand and sprinkle evenly over lawns.

Planting Your Garden

Once you've chosen a spot for your garden (as well as the size you want to make your garden bed), and prepared the soil with compost or other fertilizer, it's time to start planting. Find seeds at your local garden center, browse through seed catalogs, and order seeds that will do well in your area. Alternatively, you can start with bedding plants (or seedlings) available at nurseries and garden centers.

Read the instructions on the back of the seed package or on the plastic tag in your plant pot. You may have to ask experts when to plant the seeds if this information is not stated on the back of the package. Some seeds (such as tomatoes) should be started indoors in small pots or seed trays before the last frost, and only transplanted outdoors when the weather warms up. For established plants or seedlings, be sure to plant as directed on the plant tag or consult your local nursery about the best planting times.

Seedlings

If you live in a cooler region with a shorter growing period, you will want to start some of your plants indoors. To do this,

(continued on page 21)

SPROUTING SEEDS FOR EATING

Seeds can be sprouted and eaten on sandwiches, salads, or stir-fries any time of the year. They are delicious and full of vitamins and proteins. Mung beans, soybeans, alfalfa, wheat, corn, barley, mustard, clover, chickpeas, radishes, and lentils all make good sprouts. Find seeds for sprouting from your local health food store or use dried peas, beans, or lentils from the grocery store. Never use seeds intended for planting unless you've harvested the seeds yourself—commercially available planting seeds are often treated with a poisonous chemical fungicide.

To grow sprouts, thoroughly rinse and strain the seeds, then place in a glass jar, cover with cheesecloth secured with a rubber band, and soak overnight in cool water. You'll need about four times as much water as you have seeds. Drain the seeds by turning the jar upside down and allowing the water to escape through the cheesecloth. Keep the seeds at 60 to 80°F and rinse twice a day, draining them thoroughly after every rinse. Once sprouts are 1 to 1 ½ inches long (generally after three to five days), they are ready to eat.

RECOMMENDED PLANTS TO START AS SEEDLINGS

CROP [s] small seed [l] large seed (planting cell size)	WEEKS BEFORE TRANSPLANTING	SEED PLANTING DEPTH (Inches)	TRANSPLANT SPACING (Inches)	WITHIN ROW/ BETWEEN ROW
Broccoli [s]	1	4–6	¼–½	8–10″/18–24″
Cabbage [s]	1	4–6	¼–½	18–24″/30″
Cucumber [l]	2	4–5	½	2′/5–6′
Eggplant [s]	2	8	¼	18″/18–24″
Herbs [s]	1	4	¼	4–6″/12–18″
Lettuce [s]	2	4–5	¼	12″/12″
Melon [l]	3	4–5	¼	2–3′/6′
Onion [s]	8–12	8	¼	4″/12″
Pepper [s]	2	8	¼	12–18″/2–3′
Pumpkin [l]	3	2–4	1	5–6′/5–6′
Summer Squash [l]	3	2–4	¾–1	18″/2–3′
Tomato [s]	3	8	¼	18–24″/3′
Watermelon [l]	3	4–5	½–¾	3–4′/3–4′
Winter Squash [l]	3	2–4	1	3–4′/4–5′

Germination Temperatures of Selected Vegetable Plants			
Broccoli 77°F	Eggplant 85°F	Onion 70°F	Summer Squash 80°F
Cabbage 86°F	Herbs 65°F	Pepper 85°F	Tomato 85°F
Cucumber 86°F	Melon 90°F	Pumpkin 85°F	Winter Squash 80°F

HOMESCHOOL HINT

How do different treatments change how fast seeds sprout?

Experiment with sprouting seeds under different temperatures, or after being soaked for different times or in different liquids. Or, see how one kind of treatment affects different types of seeds. What makes seeds sprout the fastest? Do some look healthier than others? Write up a lab report to document your findings.

obtain plug flats (trays separated into many small cups or "cells") or make your own small planters by poking holes in the bottoms of paper cups. Fill the cups two-thirds full with potting soil or composted soil. Bury the seed at the recommended depth, according to the instructions on the package. Tamp down the soil lightly and water. Keep the seedlings in a warm, well-lit place, such as the kitchen, to encourage germination.

Once the weather begins to warm up and you are fairly certain you won't be getting any more frosts (you can contact your local extension office to find out the "frost free" date for your area) you can begin to acclimate your seedlings to the great outdoors. First, place them in a partially shady spot outdoors that is protected from strong wind. After a couple of days, move them into direct sunlight, and finally, transplant them to the garden.

How to Water Your Soil

After your seeds or seedlings are planted, the next step is to water your soil. Different soil types have different watering needs. You don't need to be a soil scientist to know how to water your soil properly. Here are some tips that can help to make your soil moist and primed for gardening:

1. Loosen the soil around plants so water and nutrients can be quickly absorbed.
2. Use a 1- to 2-inch protective layer of mulch on the soil surface above the root area. Cultivating and mulching help reduce evaporation and soil erosion.
3. Water your plants at the appropriate time of day. Early morning or night is the best time for watering, as evaporation is less likely to occur at these times.
4. Do not water your plants when it is extremely windy outside. Wind will prevent the water from reaching the soil where you want it to go.

Types of Soil and Their Water Retention

Knowing the type of soil you are planting in will help you best understand how to properly water and grow your garden plants. Three common types of soil and their various abilities to absorb water are listed below:

1. Clay Soil

In order to make this type of soil more loamy, add organic materials, such as compost, peat moss, and well-rotted leaves, in the spring before growing and also in the fall after harvesting your vegetables and fruits. Adding these organic materials allows this type of soil to hold more nutrients for healthy plant growth. Till or spade to help loosen the soil.

Since clay soil absorbs water very slowly, water only as fast as the soil can absorb the water.

2. Sandy Soil

As with clay soil, adding organic materials in the spring and fall will help supplement the sandy soil and promote better plant growth and water absorption.

Left on its own (with no added organic matter), the water will run through sandy soil so quickly that plants won't be able to absorb it through their roots and will fail to grow and thrive.

3. Loam Soil

This is the best kind of soil for gardening. It's a combination of sand, silt, and clay. Loam soil is fertile, deep, easily crumbles, and contains organic matter. It will help promote the growth of quality fruits and vegetables, as well as flowers and other plants.

Loam absorbs water readily and stores it for plants to use. Water as frequently as the soil needs to maintain its moisture and to promote plant growth.

Start Your Own Vegetable Garden

If you want to start your own vegetable garden, just follow these simple steps and you'll be on your way to growing your own yummy vegetables—right in your own backyard!

Steps to Making Your Own Vegetable Garden

1. Select a site for your garden.
   - Vegetables grow best in well-drained, fertile soil (loamy soils are the best).
   - Some vegetables can cope with shady conditions, but most prefer a site with a good amount of sunshine—at least six hours a day of direct sunlight.
2. Remove all weeds in your selected spot and dispose of them. If you are using compost to supplement your garden soil, do not put the weeds on the compost heap, as they may germinate once again and cause more weed growth among your vegetable plants.
3. Prepare the soil by tilling it. This will break up large soil clumps and allow you to see and remove pesky weed roots. This would also be the appropriate time to add organic materials (such as compost) to the existing soil to help make it more fertile. The tools used for tilling will depend on the size of your garden. Some examples are:
   - Shovel and turning fork—using these tools is hard work, requiring strong upper body strength.
   - Rotary tiller—this will help cut up weed roots and mix the soil.
4. After the soil has been tilled, you are ready to begin planting. If you would like straight rows in your garden, a guide can be made from two wooden stakes and a bit of rope.
5. Vegetables can be grown from seeds or transplanted:
   - If your garden has problems with pests such as slugs, it's best to transplant older plants, as they are more likely to survive attacks from these organisms.
   - Transplanting works well for vegetables like tomatoes and onions, which usually need a head start to mature within a shorter growing season. These can be germinated indoors on seed trays on a windowsill before the growing season begins.

Label your garden rows as soon as you plant the seeds or seedlings so you'll remember what you planted where.

6. Follow these basic steps to grow vegetables from seeds:
    - Information on when and how deep to plant vegetable seeds is usually printed on seed packages or on various Web sites. You can also contact your local nursery or garden center to inquire after this information.
    - Measure the width of the seed to determine how deep it should be planted. Take the width and multiply by two. That is how deep the seed should be placed in the hole. As a general rule, the larger the seed, the deeper it should be planted.
7. Water the plants and seeds well to ensure a good start. Make sure they receive water at least every other day, especially if there is no rain in the forecast.

Things to Consider

In the early days of a vegetable garden, all your plants are vulnerable to attack by insects and animals. It is best to plan multiples of the same plant in order to ensure that some survive. Placing netting and fences around your garden can help keep out certain animal pests. Coffee grains or slug traps filled with beer will also help protect your plants against insect pests.

If sowing seed straight onto your bed, be sure to obtain a photograph of what your seedlings will look like so you don't mistake the growing plant for a weed.

Weeding early on is very important to the overall success of your garden. Weeds steal water, nutrients, and light from your vegetables, which will stunt their growth and make it more difficult for them to thrive.

Growing Fruit Bushes and Trees

If you take the time to properly plan and care for your fruit bushes and trees, they'll provide you with delicious, nutrient-dense fruit year after year. Some fruit plants, like strawberries, are easy to grow and will reward you with ripe fruit relatively quickly. Fruit trees, like apple or pear, will require more work and time, but with the right maintenance they will bear fruit for generations.

Think carefully about where you choose to plant and then take time to prepare the site. Most fruit plants need at least six hours a day of sun and require well-drained soil. If the soil is not already cultivated nor relatively free of pests, spend the first year preparing the site. Planting a cover crop of rye, wheat, or oats will improve the quality of your soil and reduce weeds that could compete with your fruit plants. The cover crops will die in the late fall and add to the organic matter of the soil. Just leave them to decompose on the surface of the soil and then turn them under the soil come spring.

Testing your soil pH the year ahead of planting will give you time to adjust it if necessary to give your plants the best chance of thriving. Fruit trees, grapes, strawberries, blackberries, and raspberries do best if the soil pH is between 6.0 and 6.5. Blueberries require a more acidic soil, around 4.5.

What plants will thrive will depend largely on where you live, your planting zone (see page 3), your altitude, and your proximity to large bodies of water (since areas close to water tend to be more temperate). Refer to the chart below for hardiness zones for most fruits, but keep in mind that hardiness varies by variety. Refer to seed catalogs or talk to other local gardeners before settling on a particular variety of fruit to plant.

Strawberries

Purchase young strawberry plants to plant in the spring after the last frost. Try to find plants that are certified disease-free, since diseases from strawberries can spread through your whole garden. Strawberries thrive with lots of sun and well-drained soil. If you have access to a gentle south-facing slope, this is ideal. Till the top 12 inches of soil. If you planted a cover crop, turn under all the organic matter. If not, be sure to add manure or compost to reach a rich, slightly acidic soil.

FRUIT GROWING CHART

Fruit	Hardiness Zone	Soil pH	Space Between Plants (in Inches)	Space Between Rows (in Inches)	Bearing Age (in Years)	Potential Yield (in Pounds)	When to Harvest
Apple	5 to 7	6 to 6.5	7 to 18 (depending on variety)	13 to 24 (depending on variety)	3 to 5	60 to 250	August to October
Apricot	4 to 8	6 to 6.5	15	20	4	100	July to August
Blackberry	3 to 9 (depending on variety)	6 to 6.5	2	10	2	2 to 3	July to August
Blueberry	3 to 11 (depending on variety)	4 to 5	4 to 5	10	3 to 6	3 to 10	July to September
Cherry, sweet	5 to 7	6 to 6.5	24	30	7	300	July
Cherry, tart	4 to 7	6 to 6.5	18	24	4	100	July
Currant	2 to 6	5.5 to 7	4	8	2 to 4	6 to 8	July
Elderberry	3 to 9	6 to 6.5	6	10	2 to 4	4 to 8	August to September
Gooseberry	2 to 6	5.5 to 7	4	10	2 to 4	2 to 4	July to August
Grapes	5 to 10 (depending on variety)	6 to 6.5	8	9	3	10 to 20	September to October
Peach and nectarine	5 to 8	6 to 6.5	15	20	4 to 5	100	August to September
Pear and quince	4 to 7	6 to 6.5	15 to 20	15 to 20	4	100	August to October
Plum	5 to 7	6 to 6.5	10	15	5	75	July to September
Raspberry	3 to 8	5.6 to 6.2	2	8	2	1 to 2	July to September
Strawberry	4 to 9	5.5 to 6.5	12 to 18	12 to 18	1 to 2	1 to 3	May to July

Dig a 5- to 7-inchwide hole for each plant. It should be deep enough to accommodate the root system without squishing it. Place the plant in the hole and fill in the soil, tamping it down gently around the plant. Space plants about 12 inches apart on all sides. The roots will shoot out runners that produce more small plants. To allow the plants to focus their energy on fruit production, snip the runners and transplant or discard any new plants.

An alternate planting method is the matted-row system. This method requires less maintenance but offers a slightly lower quality yield. Space plants about 18 inches apart, allowing the roots to shoot out runners and produce new plants. If planting more than one row, space them three to four feet apart. To aid picking and to keep the plants from competing with each other, prune out the plants on the outer edges of rows by snipping the runners and pulling out the plants.

Strawberry plants should receive at least an inch of water per week. In the first year, snip away or pick blossoms as soon as they develop. You will sacrifice your fruit crop in the first year, but you will have healthier plants and a greater fruit yield for many years afterward.

Brambles and Bush Fruits

Most brambles and bush fruits should be planted in the spring after the last frost. Blackberries and raspberries should have any old or damaged canes removed before planting in a 4-inch-deep hole or furrow. Do not fertilize for several weeks after planting and even after that use fertilizer sparingly; brambles are easily damaged by overfertilizing. If the weather is dry, water bushes in the morning, just after the dew has dried, being careful to avoid getting the foliage very wet.

Brambles need to be pruned once a year to keep them healthy and to keep them from spreading out of control. How you prune your brambles will depend on the variety. Fall-bearing brambles (primocane-fruiting brambles) produce fruit the same year they are planted. If planted in spring, they'll produce some berries in late summer or early fall and then again (lower on the canes) in early summer of their second year. For these varieties, there are two pruning choices. The first option produces a smaller yield but higher quality berries in late summer. For this pruning method, mow down the bush all the way to the ground in late fall. The second method will produce berries in summer and fall, and is the same method used for floricane-fruiting brambles. Allow the canes to grow through the first year and prune them gently in the early spring of the following year. Remove any damaged or diseased parts and thin canes to three or four per foot. Trim down the tops of canes to about 4 to 5 feet high so that you can easily pick the berries. Prune similarly every spring.

Blueberries

Blueberries thrive in acidic soil (around pH 4.8). If your soil is naturally over pH 6.5, you're better off planting your blueberries in container gardens or raised beds, where you can more easily manipulate the soil's pH.

Blueberries should be planted in a hole about 6 inches deep and 20 inches in diameter. The crowns should be right at soil level. Surround the stems with about 6 inches of sawdust mulch or leaves and water the bushes in the morning in dryer climates. Blueberries are very sensitive to drought.

Some varieties of blueberries only require pruning if there are dead or damaged branches that need to be removed. Highbush varieties should be pruned every year starting after the third year after planting. When old canes get twiggy, cut them off at soil level to allow new, stronger canes to emerge.

Currants and Gooseberries

These berries thrive in colder regions in well-drained soils. Plant them in holes 12 inches across and 12 inches deep. They should bear fruit in their second year and will give the highest yield in their third year. After that, the canes will begin to darken, which is a sign that it's time to prune them by mowing or clipping at soil level. By this point, new canes will likely have developed. Currants and gooseberries do not require much pruning, but dead or diseased branches should always be removed.

Fruit Trees

Planting

Once you've decided on which varieties to plant and have planned and prepared the best site, it's time to purchase the trees and plant them. Most young trees come with the roots planted in a container of soil, embedded in a ball of soil and wrapped in burlap (balled-and-burlapped, or B&B), or packed in damp moss or excelsior. It's best to plant your trees as soon as possible after purchasing, though B&B stock or potted trees can be kept for several weeks in a shady area.

To plant, dig a hole that is about 2 feet deep and wide enough to give the roots plenty of room to spread (about 1 ½ feet wide). As you dig, try to keep the sod, topsoil, and subsoil in separate piles. Once the hole is the right size, loosen the dirt at the bottom and then place the sod into the hole, upside down. Then make a small mountain of topsoil in the center of the hole. The roots will sit on top of the mountain and hang over the edges.

Next, prune away any broken or damaged roots. Pruning shears or a sharp knife work best. Remove roots that are very tangled or too long to fit in the hole. Place the roots into the hole, on top of the dirt mound. If the tree was grafted, it will have a "graft union," a bulge where the roots meet the trunk. For most trees, this mound should be barely visible above the ground. For dwarf trees, it should be just below the soil level.

Most young trees need extra support. Drive a 5- to 6-foot garden stake into the ground a few inches away from the trunk and on the south side. The stake should go about 2 feet deeper than the roots.

Begin filling the hole back in with soil, using your fingers to work the soil around the roots and eliminate any air pockets. Add soil until the hole is filled, pat it down until it's slightly lower than the ground level (to help retain water), and then pour a bucket of water over the soil to pack it down further. Prune away all but the three or four strongest branches and then tie the trunk gently to the stake with a soft rag.

Spread leaves, mulch, or bark around the base of the tree to protect the roots and help retain the water. Because young tree bark is easily injured, wrap the trunk carefully with burlap from the ground to its lowest branches. This will protect the bark from being scalded by sun (even in winter) or damaged by deer or rodents. Leave the wrap on for the first two or three years.

Pruning

Your fruit trees will benefit from gentle pruning once a year. The goals of pruning are to remove dead or damaged branches, to keep branches from

» Blackcurrants do best planted in full sun and moist, slightly acidic (pH 6 to 6.5) soil. The berries are tart and perfect for use in jams, pies, and other desserts.

crowding each other, and to keep trees from growing so large that they begin to invade each other's space. Pruning, when done properly, will produce healthier trees and more fruit.

Pruning shears or a pruning saw can be used. When removing a whole branch, try to cut flush with the trunk, so that no "stub" is left behind. Stubs soon decay, inviting insects to invade your tree. A cut that is flush with the trunk will heal over quickly (in one growing season). If you're removing part of a branch, cut slightly above a bud and cut at a slant. Try to choose a bud that slants in the direction you want a new branch to grow.

When removing particularly large branches, care should be taken to ensure that the branch doesn't tear away large pieces of bark from the trunk as it falls. To do this, start below the branch and cut upwards about a third of the way through the branch. Then cut down from the top, starting an inch or two further away from the trunk.

If your pruning leaves a wound that is larger than a silver dollar, use a knife to peel away the bark above and below the wound to create a vertical diamond shape. Then cover the wound with shellac or tree wound paint to protect it from decay and insects.

Grapes

Talk to someone at your local nursery or to other growers in your area to determine which grape variety will work best in your location. All varieties fall under the categories of wine, table, or slipskin. Grapes need full sun to stay healthy, and benefit from loose, well-drained, loamy soil.

Before planting your vine cutting, soak it in a bucket of water for at least six hours. Cuttings should never dry out. Grape vines need a trellis or a similar support. Dig a hole near the trellis deep enough to accommodate the roots and douse the hole with water. Place the cutting in the hole and fill in the soil around it, adding more water as you go, and then tamping it down firmly. Prune away all but the best cane and tie it gently to the support (a stake or the bottom wire of a trellis). Water the vine once a week for at least the first month.

Don't use any fertilizer in the first year as it can actually damage the young vines. If necessary, begin fertilizing the soil in the second year.

Buds will begin to grow after several weeks. After about ten weeks, remove all but the strongest shoots as well as any flower clusters or side shoots. Every year in the late fall, after the last grapes have been picked, remove 90 percent of the new growth from that season. You should be able to harvest fruit in the plant's third year.

A simple grape arbor can be made by firmly lashing or nailing together trees with their bark removed. Be sure the supporting posts are inserted at least a few feet into the ground.

Growing and Threshing Grains

Grains are a type of grass and they grow almost as easily as the grass in your yard does. There are many reasons for growing your own grains, including supplying feed for your livestock, providing food for you and your family, or to use as a green manure (a crop that will be plowed back into the soil to enrich it). Growing grains requires much less work than growing a vegetable garden, though getting the grains from the field to the table requires a bit more work.

Whether you are growing wheat, oats, barley, or another grain, the process is basically the same:

1. Decide which grain to grow. Most cereal grains have a spring variety and a winter variety. Winter grains are often preferred because they are more nutritious than spring varieties and are less affected by weeds in the spring. However, spring wheat is preferred in cold climates as winter wheat may not survive very harsh winters. If you have trouble finding smaller amounts of seeds to purchase from seed supply houses, try health food stores. They often have bins full of grains you can buy in bulk for eating, and they work just as well for planting, as long as you know what variety of grain you're buying. Winter grains

should be planted from late September to mid-October, after most insects have disappeared but before the hard frosts set in. Spring wheat should be planted in early spring.

2. Decide how much grain you want to grow. A 10-foot by 10-foot plot of wheat will provide enough flour for about twenty loaves of bread. An acre of corn will provide feed for a pig, a milk cow, a beef steer, and thirty laying hens for an entire year.
3. Prepare the soil. Rototill or use a shovel to turn over the earth, remove any stones or weeds, and make the plot as even as possible using a garden rake.
4. Sprinkle the seeds over the entire plot. How much seed you use will depend on the grain (refer to the chart on page 32). For wheat, use a ratio of around 3 ounces of seed per 100 square feet. Aim to plant about one seed per square inch. Rake over the plot to cover all the seeds with earth.
5. Water the seeds immediately after planting and then about once a month throughout the growing season if there's not adequate rainfall.
6. When the grain is golden with a few streaks of green left, it's ready for harvest. For winter grains, harvest is usually ready in June or July. To cut the grains, use a scythe, machete, or other sharp knife, and cut near the base of the stems. Gather the grains into bundles, tie them with twine, and stand them upright in the plot to finish ripening. Lean three or four bundles together to keep them from falling over. If there is danger of rain, move the sheaves into a barn or other covered area to prevent them from molding. Once all the green has turned to gold, the grains are ready for threshing.
7. The simplest way to thresh is to grasp a bunch of stalks and beat it around the inside of a barrel, heads facing down. The grain will fall right off the stalks. Alternatively, you can lay the stalks down on a hard surface covered by an old sheet and beat the seed heads with a broom or baseball bat. Discard (or compost) the stalks. If there is enough breeze, the chaff will blow away, leaving only the grains. You can also pour the grain and chaff back and forth between two barrels and allow the wind (which can be supplied by a fan if necessary) to blow away the chaff.
8. Store grain in a covered metal trash can or a wooden bin. Be sure it is kept completely dry and that no rodents can get in.

HOW MUCH GRAIN SHOULD YOU GROW TO FEED YOUR FAMILY?

An acre of wheat will supply about 30 bushels of grain, or around 1,800 pounds. The average American consumes about 140 pounds of wheat in a year. The Federal Emergency Management Association (FEMA) recommends the following consumption rates:

- Adult males, pregnant or nursing mothers, active teens ages 14 to 18: 275 lbs./year
- Women, kids ages 7 to 13, seniors: 175 lbs./year
- Children 6 and under: 60 lbs./year

GRAIN GROWING CHART

Type of Grain	Amount of seed per acre (in pounds)	Grain yield per acre (in bushels)	Characteristics and Uses
Amaranth	1	125	Very tolerant of arid environments. High in protein and gluten-free. Use in baking or animal feed.
Barley	100	120 to 140	Tolerates salty and alkaline soils better than most grains. Use in animal feed, soups, as a side dish, and for making beer and malts.
Buckwheat	50	20 to 30	Matures rapidly (sixty to ninety days). Rich, nutty flavor perfect for baking and in pancakes.
Field corn	6 to 8	180 to 190	Use in animal feed, corn starch, hominy, and grits.
Grain sorghum	2 to 8	70 to 100	Drought tolerant. Use in animal feed or in baking.
Oats	80	70 to 100	Thrives in cool, moist climates. High in protein. Use in animal feed, baking, or as a breakfast cereal.
Rye	84	25 to 30	Tolerant of cold, dampness, and drought. Use for animal feed, in baking, to make whiskey, or as a cover crop.
Wheat	75 to 90	40 to 70	Hard red winter wheat is used in bread and is highly nutritious. Soft red winter wheat is good for cakes and pastries. Hard red spring wheat is the most common bread wheat. Durum wheat is best for pasta.

Container Gardening

An alternative to growing vegetables, flowers, and herbs in a traditional garden is to grow them in containers. While the amount that can be grown in a container is certainly limited, container gardens work well for tomatoes, peppers, cucumbers, herbs, salad greens, and many flowering annuals. Choose vegetable varieties that have been specifically bred for container growing. You can obtain this information online or at your garden center. Container gardening also brings birds and butterflies right to your doorstep. Hanging baskets of fuchsia or pots of snapdragons are frequently visited by hummingbirds, allowing for up-close observation.

Container gardening is an excellent method of growing vegetables, herbs, and flowers, especially if you do not have adequate outdoor space for a full garden bed. A container garden can be placed anywhere—on the patio, balcony, rooftop, or windowsill. Vegetables such as leaf lettuce, radishes, small tomatoes, and baby carrots can all be grown successfully in pots.

How to Grow Vegetables in a Container

Here are some simple steps to follow for growing vegetables in containers:

1. Choose a sunny area for your container plants. Your plants will need at least five to six hours of sunlight a day. Some plants, such as cucumbers, may need more. Select plants that are suitable for container growing. Usually their names will contain words such as "patio," "bush," "dwarf," "toy," or "miniature." Peppers, onions, and carrots are also good choices.
2. Choose a planter that is at least five gallons, unless the plant is very small. Poke holes in the bottom if they don't already exist; the soil must be able to drain in order to prevent the roots from rotting. Avoid terra-

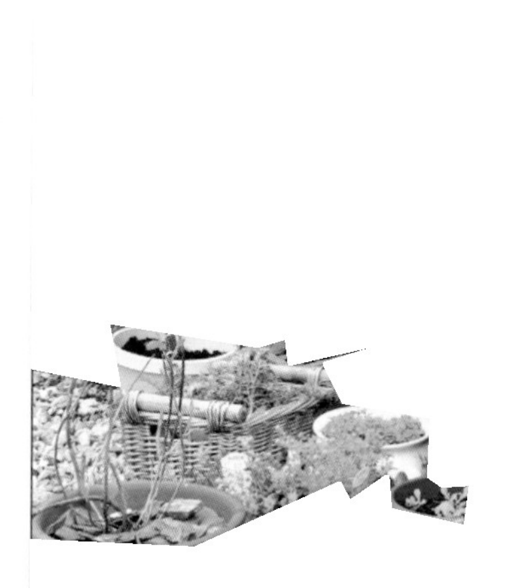

cotta or dark colored pots as they tend to dry out quickly.

3. Fill your container with potting soil. Good potting soil will have a mixture of peat moss and vermiculite. You can make your own potting soil using composted soil. Read the directions on the seed packet or label to determine how deep to plant your seeds.
4. Check the moisture of the soil frequently. You don't want the soil to become muddy, but the soil should always feel damp to the touch. Do not wait until the plant is wilting to water it—at that point, it may be too late.

Things to Consider

- Follow normal planting schedules for your climate when determining when to plant your container garden.
- You may wish to line your container with porous materials such as shredded newspaper or rags to keep the soil from washing out. Be sure the water can still drain easily.

How to Grow Herbs in a Container

Herbs will thrive in containers if cared for properly. And if you keep them near your kitchen, you can easily snip off pieces to use in cooking. Here's how to start your own herb container garden:

1. If your container doesn't already have holes in the bottom, poke several to allow the soil to drain. Pour gravel into the container until it is about a quarter of the way full. This will help the water drain and help to keep the soil from washing out.
2. Fill your container three-quarters of the way with potting soil or a soil-based compost.
3. It's best to use seedlings when planting herbs in containers. Tease the roots slightly, gently spreading them apart with your fingertips. This will encourage them to spread once planted. Place each herb into the pot and cover the root base with soil. Place herbs that will grow taller in the center of your container, and the

smaller ones around the edges. Leave about 4 square inches of space between each seedling.

4. As you gently press in soil between the plants, leave an inch or so between the container's top and the soil. You don't want the container to overflow when you water the herbs.
5. Cut the tops off the taller herb plants to encourage them to grow faster and to produce more leaves.
6. Pour water into the container until it begins to leak out the bottom. Most herbs like to dry out between watering, and overwatering can cause some herbs to rot and die, so only water every few days unless the plants are in a very hot place.

Things to Consider

- Growing several kinds of herbs together helps the plants to thrive. A few exceptions to this rule are oregano, lemon balm, and tea balm. These herbs should be planted on their own because they will overtake the other herbs in your container.
- You may wish to choose your herbs according to color to create attractive arrangements for your home. Any of the following herbs will grow well in containers:
  - Silver herbs: artemisias, curry plants, santolinas
  - Golden herbs: lemon thyme, calendula, nasturtium, sage, lemon balm
  - Blue herbs: borage, hyssop, rosemary, catnip
  - Green herbs: basil, mint, marjoram, thyme, parsley, chives, tarragon
  - Pink and purple herbs: oregano (the flowers are pink), lavender
- If you decide to transplant your herbs in the summer months, they will grow quite well outdoors and will give you a larger harvest.

How to Grow Flowers from Seeds in a Container

1. Cover the drainage hole in the bottom of the pot with a flat stone. This will keep the soil from trickling out when the plant is watered.
2. Fill the container with soil. The container should be filled almost to the top. For the best results, use potting soil from your local nursery or garden center.
3. Make holes for the seeds. Refer to the seed packet to see how deep to make the holes. Always save the seed packet for future reference—it most likely has helpful directions about thinning young plants.
4. Place a seed in each hole. Pat the soil gently on top of each seed.
5. Use a light mist to water your seeds, making sure that the soil is only moist and not soaked.
6. Make sure your seeds get the correct amount of sunlight. Refer to the seed packet for the adequate amount of sunlight each seedling needs.
7. Watch your seeds grow. Most seeds take three to seventeen days to sprout. Once the plants start sprouting, be sure to pull out plants that are too close together so the remaining plants will have enough space to establish good root systems.
8. Remember to water and feed your container plants. Keep the soil moist so your plants can grow. And in no time at all, you should have wonderful flowers growing in your container garden.

Preserving Your Container Plants

As fall approaches, frost will soon descend on your container plants and can ultimately destroy your garden. Container plants are particularly susceptible to frost damage, especially if you are growing tropical plants, perennials, and hardy woody plants in a single container garden. There are many ways that you can preserve and maintain your container garden plants throughout the winter season.

Preservation techniques will vary depending on the plants in your container garden. Tropical plants can be overwintered using methods replicating a dry season, forcing the plant into dormancy; hardy perennials and woody shrubs need a cold dormancy to grow in the spring, so they must stay outside; cacti and succulents prefer their winters warm and dry and must be brought inside; while many annuals can be propagated by stem cuttings or can just be repotted and maintained inside.

Preserving Tropical Bulbs and Tubers

Many tropical plants, such as cannas, elephant ears, and angel's trumpets can be saved from an untimely death by overwintering them in a dark corner or sunny window of your home, depending on the type of plant. A lot of bulbous and tuberous tropical plants have a natural dry season (analogous to our winter) when their leafy parts die off, leaving the bulb behind. Don't throw the bulbs away. After heavy frosts turn the aboveground plant parts to mush, cut the damaged foliage off about 4 inches above the thickened bulb. Then, dig them up and remove all excess soil from the roots. If a bulb has been planted for several years and its performance is beginning to decline, it may need dividing. Daffodils, for example, should

generally be divided every three years. If you do divide the bulb, be sure to dust all cut surfaces with a sulfur-based fungicide made for bulbs to prevent the wounds from rotting. Cut the roots back to 1 inch from the bulb and leave to dry out evenly. Rotten bulbs or roots need to be thrown away so infection doesn't spread to the healthy bulbs.

A bulb's or tuber's drying time can last up to two weeks if it is sitting on something absorbent like newspaper and located somewhere shaded and dry, such as a garage or basement. Once clean and dry, bulbs should be stored—preferably at around 50°F—all winter in damp (not soggy) milled peat moss. This prevents the bulbs from drying out any further, which could cause them to die. Many gardeners don't have a perfectly cool basement or garage to keep bulbs dormant. Alternative methods for dry storage include a dark closet with the door cracked for circulation, a cabinet, or underneath a bed in a cardboard box with a few holes punched for airflow. The important thing to keep in mind is that the bulb needs to be kept on the dry side, in the dark, and moderately warm.

If a bulb was grown as a single specimen in its own pot, the entire pot can be placed in a garage that stays above 50°F or a cool basement and allowed to dry out completely. Cut all aboveground plant parts flush with the soil and don't water until the outside temperatures stabilize above 60°F. Often, bulbs break dormancy unexpectedly in this dry pot method. If this happens, pots can be moved to a sunny location near a window and watered sparingly until they can be placed outside. The emerging leaves will be stunted, but once outside, the plant will replace any spindly leaves with lush, new ones.

Annuals

Many herbaceous annuals can also be saved for the following year. By rooting stem cuttings in water on a sunny windowsill, plants like impatiens, coleus, sweet potato vine cultivars, and purple heart can be held over winter until needed in the spring. Otherwise, the plants can be cut back by half, potted in a peat-based, soilless mix, and placed on a sunny windowsill. With a wide assortment of "annuals" available on the market, some research is required to determine which annuals can be overwintered successfully. True annuals (such as basils, cockscomb, and zinnias)—regardless of any treatment given—will go to seed and die when brought inside.

Cacti and Succulents

If you planted a mixed dry container this year and want to retain any of the plants for next year, they should be removed from the main container and repotted into a high sand-content soil mix for cacti and succulents. Keep them in a sunny window and water when dry. Many succulents and cacti do well indoors, either in a heated garage or a moderately sunny corner of a living room.

As with other tropical plants, succulents also need time to adjust to sunnier conditions in the spring. Move them to a shady spot outside when temperatures have stabilized above 60°F and then gradually introduce them to brighter conditions.

Hardy Perennials, Shrubs, and Vines

Hardy perennials, woody shrubs, and vines needn't be thrown away when it's time to get rid of accent containers. Crack-resistant, four-season containers can house perennials and woody shrubs year-round. Below is a list of specific perennials and woody plants that do well in both hot and cold weather, indoors and out:

- Shade perennials, like coral bells, lenten rose, assorted hardy ferns, and Japanese forest grass are great for all weather containers.
- Sun-loving perennials, such as sedges, some salvias, purple coneflower, daylily, spiderwort, and bee blossom are also very hardy and do well in year-round containers. Interplant them with cool growing plants, like kale, pansies, and Swiss chard, for fall and spring interest.
- Woody shrubs and vines—many of which have great foliage interest with four-season appeal—are ideal for container gardens. Red-twigged dogwood cultivars, clematis vine cultivars, and dwarf crape myrtle cultivars are great container additions that can stay outdoors year-round.

If the container has to be removed, hardy perennials and woody shrubs can be temporarily planted in the ground and mulched. Dig them from the garden in the spring, if you wish, and replant into a container. Or, leave them in their garden spot and start over with fresh ideas and new plant material for your container garden.

Sustainable Plants and Money in Your Pocket

Overwintering is a great form of sustainable plant conservation achieved simply and effectively by adhering to each plant's cultural and environmental needs. With careful planning and storage techniques, you'll save money as well as plant material. The beauty and interest you've created in this season's well-grown container garden can also provide enjoyment for years to come.

Raised Beds

If you live in an area where the soil is quite wet (preventing a good vegetable garden from growing in the spring), or find it difficult to bend over to plant and cultivate your vegetables or flowers, or if you just want a different look to your backyard garden, you should consider building a raised bed.

A raised bed is an interesting and affordable way to garden. It creates an ideal environment for growing vegetables, since the soil concentration can be closely monitored and, as it is raised above the ground, it reduces the compaction of plants from people walking on the soil.

Raised beds are typically 2 to 6 feet wide and as long as needed. In most cases, a raised bed consists of a "frame" that is filled in with nutrient-rich soil (including compost or organic fertilizers) and is then planted with a variety of vegetables or flowers, depending on the gardener's preference. By controlling the bed's construction and the soil mixture that goes into the bed, a gardener can effectively reduce the amount of weeds that will grow in the garden.

When planting seeds or young sprouts in a raised bed, it is best to space the plants equally from each other on all sides. This will ensure that the leaves will be touching once the plant is mature, thus saving space and reducing the soil's moisture loss.

How to Make a Raised Bed

Step One: Plan Out Your Raised Bed

1. Think about how you'd like your raised bed to look, and then design the shape. A raised bed is not extremely complicated, and all you need to do is build an open-top and open-bottom box (if you are ambitious, you can create a raised bed in the shape of a circle, hexagon, or star). The main purpose of this box is to hold soil.
2. Make a drawing of your raised bed, measure your available garden space, and add those measurements to your drawing. This will allow you to determine how much material is needed. Generally, your bed should be at least 24 inches in height.
3. Decide what kind of material you want to use for your raised bed. You can use lumber, plastic, synthetic wood, railroad ties, bricks, rocks, or a number of other items to hold the dirt. Using lumber is the easiest and most efficient method.
4. Gather your supplies.

Step Two: Build Your Raised Bed

1. Make sure your bed will be situated in a place that gets plenty of sunlight. Carefully assess your placement, as your raised bed will be fairly permanent.
2. Connect the sides of your bed together (with either screws or nails) to form the desired shape of your bed. If you are using lumber, you can use 4-inch x 4-inch posts to serve as the corners of your bed, and then nail or screw the sides to these corner posts. By doing so, you will increase strength of the structure and ensure that the dirt will stay inside.
3. Cut a piece of gardening plastic to fit inside your raised bed. This will significantly reduce the amount of weeds growing in your garden. Lay it out in the appropriate location.
4. Place your frame over the gardening plastic (this might take two people).

Step Three: Start Planting

1. Add some compost into the bottom of the bed and then layer potting soil on top of the compost. If you have soil from other parts of your yard, feel free to use that in addition to the compost and potting soil. Plan on filling at least one-third of your raised bed with compost or composted manure (available from nurseries or garden centers in 40-pound bags).
2. Mix in dry organic fertilizers (like wood ash, bone meal, and blood meal) while building your bed. Follow the package instructions for how best to mix it in.
3. Decide what you want to plant. Some people like to grow flowers in their raised beds; others prefer to grow vegetables. If you do want to grow food, raised beds are excellent choices for

THINGS YOU'LL NEED

- Forms for your raised bed (consider using 4-inch x 4-inch posts cut to 24 inches in height for corners, and 2-inch x 12-inch boards for the sides)
- Nails or screws
- Hammer or screwdriver
- Plastic liner (to act as a weed barrier at the bottom of your bed)
- Shovel
- Compost or composted manure
- Soil (either potting soil or soil from another part of your yard)
- Rake (to smooth out the soil once in the bed)
- Seeds or young plants
- Optional: PVC piping and greenhouse plastic (to convert your raised bed to a greenhouse)

salad greens, carrots, onions, radishes, beets, and other root crops.

Things to Consider

1. To save money, you can dig up and use soil from your yard. Potting soil can be expensive, and yard soil is just as effective when mixed with compost. However, removing grass and weeds from the soil before filling your raised beds can be time-consuming.
2. Be creative when building your raised planting bed. You can construct a great raised bed out of recycled goods or old lumber.
3. You can convert your raised bed into a greenhouse. Just add hoops to your bed by bending and connecting PVC pipe over the bed. Then clip greenhouse plastic to the PVC pipes, and you have your own greenhouse.
4. Make sure to water your raised bed often. Because it is above ground, your raised bed will not retain water as well as the soil in the ground. If you keep your bed narrow, it will help conserve water.
5. Decorate or illuminate your raised bed to make it a focal point in your yard.
6. If you use lumber to construct your raised bed, keep a watch out for termites.
7. Beware of old pressure-treated lumber, as it may contain arsenic and could potentially leak into the root systems of any vegetables you might grow in your raised bed. Newer pressure-treated lumber should not contain these toxic chemicals.

Pest and Disease Management

Pest management can be one of the greatest challenges to the home gardener. Yard pests include weeds, insects, diseases, and some species of wildlife. Weeds are plants that are growing out of place. Insect pests include an enormous number of species from tiny thrips that are nearly invisible to the naked eye, to the large larvae of the tomato hornworm. Plant diseases are caused by fungi, bacteria, viruses, and other organisms—some of which are only now being classified. Poor plant nutrition and misuse of pesticides also can cause injury to plants. Slugs, mites, and many species of wildlife, such as rabbits, deer, and crows can be extremely destructive as well.

Identify the Problem

Careful identification of the problem is essential before taking measures to control the issue in your garden. Some insect damage may at first appear to be a disease, especially if no visible insects are present. Nutrient problems may also mimic diseases. Herbicide damage, resulting from misapplication of chemicals, can also be mistaken for other problems. Learning about different types of garden pests is the first step in keeping your plants healthy and productive.

Insects and Mites

All insects have six legs, but, other than that, they are extremely different depending on the species. Some insects include such organisms as beetles, flies, bees, ants, moths, and butterflies. Mites and spiders have eight legs—they are not, in fact, insects but will be treated as such for the purposes of this section.

Insects damage plants in several ways. The most visible damage caused by insects is chewed plant leaves and flowers. Many pests are visible and can be readily identified, including the Japanese beetle, Colorado potato beetle, and numerous species of caterpillars such as tent caterpillars and tomato hornworms.

Japanese beetle

Other chewing insects, however, such as cutworms (which are caterpillars) come out at night to eat, and burrow into the soil during the day. These are much harder to identify but should be considered likely culprits if young plants seem to disappear overnight or are found cut off at ground level.

Sucking insects are extremely common in gardens and can be very damaging to your vegetable plants and flowers. The most known of these insects are leafhoppers, aphids, mealy bugs, thrips, and mites. These insects insert their mouthparts into the plant tissues and suck out the plant juices. They also may carry diseases that they spread from plant to plant as they move about the yard. You may suspect that these insects are present if you notice misshapen plant leaves or flower petals. Often the younger leaves will appear curled or puckered. Flowers developing from the buds may only partially develop if they've been sucked by these bugs. Look on the undersides of the leaves—that is where many insects tend to gather.

Other insects cause damage to plants by boring into stems, fruits, and leaves, possibly disrupting the plant's ability to transport water. They also create opportunities for disease organisms to attack the plants. You may suspect the presence of boring insects if you see small accumulations of sawdust-like material on plant stems or fruits. Common examples of boring insects include squash vine borers and corn borers.

Integrated Pest Management (IPM)

It is difficult, if not impossible, to prevent all pest problems in your garden every year. If your best prevention efforts have not been entirely successful, you may need to use some control methods. Integrated pest management (IPM) relies on several techniques to keep pests at acceptable population levels without excessive use of chemical controls. The basic principles of IPM include monitoring (scouting), determining tolerable injury levels (thresholds), and applying appropriate strategies and tactics to solve the pest issue. Unlike other methods of pest control where pesticides are applied on a rigid schedule, IPM applies only those controls that are needed, when they are needed, to control pests that will cause more than a tolerable level of damage to the plant.

Monitoring

Monitoring is essential for a successful IPM program. Check your plants regularly. Look for signs of damage from insects and diseases as well as indications of adequate fertility and moisture. Early identification of potential problems is essential.

There are thousands of insects in a garden, many of which are harmless or even beneficial to the plants. Proper identification is needed before control strategies can be adopted. It is important to recognize the different stages of insect development for several reasons. The caterpillars eating your plants may be the larvae of the butterflies you were trying to attract. Any small larva with six spots on its back is probably a young ladybug, a very beneficial insect.

Thresholds

It is not necessary to kill every insect, weed, or disease organism invading your garden in order to maintain the plants' health. When dealing with garden pests, an economic threshold comes into play and is the point where the damage caused by the pest exceeds the cost of control. In a home garden, this can be difficult to determine. What you are growing and how you intend to use it will determine how much damage you are willing to tolerate. Remember that larger plants, especially those close to harvest, can tolerate more damage than a tiny seedling. A few flea beetles on a radish seedling may warrant control, whereas numerous Japanese beetles eating the leaves of beans close to harvest may not.

If the threshold level for control has been exceeded, you may need to employ control strategies. Effective and safe strategies can be discussed with your local Cooperative Extension Service, garden centers, or nurseries.

Mechanical/Physical Control Strategies

Many insects can simply be removed by hand. This method is definitely preferable if only a few, large insects are causing the problem. Simply remove the insect from the plant and drop it into a container of soapy water or vegetable oil. Be aware that some insects have prickly spines or excrete oily substances that can cause injury to humans. Use caution when handling unfamiliar insects. Wear gloves or remove insects with tweezers.

Many insects can be removed from plants by spraying water from a hose or sprayer. Small vacuums can also be used to suck up insects. Traps can be used effectively for some insects as well. These come in a variety of styles depending on the insect to be caught. Many traps rely on the use of pheromones—naturally occurring chemicals produced by the insects and used to attract the opposite sex during mating. They are extremely specific for each species and, therefore, will not harm beneficial species. One caution with traps is that they may actually draw more insects into your yard, so don't place them directly into your garden. Other traps (such as yellow and blue sticky cards) are more generic and will attract numerous species. Different insects are attracted to different colors of these traps. Sticky cards also can be used effectively to monitor insect pests.

Other Pest Controls

Diatomaceous earth, a powder-like dust made of tiny marine organisms called diatoms, can be used to reduce damage from soft-bodied insects and slugs. Spread this material on the soil—it is sharp and cuts or irritates these soft organisms. It is harmless to other organisms. In order to trap slugs, put out shallow dishes of beer.

Biological Controls

Biological controls are nature's way of regulating pest populations. Biological controls rely on predators and parasites to keep organisms under control. Many of our present pest problems result from the loss of predator species and other biological control factors.

Some biological controls include birds and bats that eat insects. A single bat can eat up to 600 mosquitoes an hour. Many bird species eat insect pests on trees and in the garden.

Chemical Controls

When using biological controls, be very careful with pesticides. Most common pesticides are broad spectrum, which means that they kill a wide variety of organisms. Spray applications of insecticides are likely to kill numerous beneficial insects as well as the pests. Herbicides applied to weed species may drift in the wind or vaporize in the heat of the day and injure non-targeted plants. Runoff of pesticides can pollute water. Many pesticides are toxic to humans as well as pets and small animals that may enter your yard. Try to avoid using these types of pesticides at all costs—and if you do use them, read the labels carefully and avoid spraying them on windy days.

Some common, non-toxic household substances are as effective as many toxic pesticides. A few drops of dishwashing detergent mixed with water and sprayed on plants is extremely effective in controlling many soft-bodied insects, such as aphids and whiteflies.

A worm-eaten apple vs. a healthy apple

Beneficial Insects that Help Control Pest Populations

Insect	Pest Controlled
Green lacewings	Aphids, mealy bugs, thrips, and spider mites
Ground beetles	Caterpillars that attack trees and shrubs
Ladybugs	Aphids and Colorado potato beetles
Praying mantises	Almost any insect
Seedhead weevils and other beetles	Weeds

Crushed garlic mixed with water may control certain insects. A baking soda solution has been shown to help control some fungal diseases on roses.

Alternatives to Pesticides and Chemicals

When used incorrectly, pesticides can pollute water. They also kill beneficial as well as harmful insects. Natural alternatives prevent both of these events from occurring and save you money. Consider using natural alternatives to chemical pesticides: Non-detergent insecticidal soaps, garlic, hot pepper spray, 1 teaspoon of liquid soap in a gallon of water, used dishwater, or a forceful stream of water from a hose all work to dislodge insects from your garden plants.

Another solution is to consider using plants that naturally repel insects. These plants have their own chemical defense systems and, when planted among flowers and vegetables, they help keep unwanted insects away.

Plant Diseases

Plant disease identification is extremely difficult. In some cases, only laboratory analysis can conclusively identify some diseases. Disease organisms injure plants in several ways: Some attack leaf surfaces and limit the plant's ability to carry on photosynthesis; others produce substances that clog plant tissues that transport water and nutrients; still other disease organisms produce toxins that kill the plant or replace plant tissue with their own.

Symptoms that are associated with plant diseases may include the presence of mushroom-like growths on trunks of trees; leaves with a grayish, mildewed appearance; spots on leaves, flowers, and fruits; sudden wilting or death of a plant or branch; sap exuding from branches or trunks of trees; and stunted growth.

Misapplication of pesticides and nutrients, air pollutants, and other environmental conditions—such as flooding and freezing—can

» Aphids

» Cutworms

Natural Pest Repellants

Pest	Repellant
Ant	Mint, tansy, or pennyroyal
Aphids	Mint, garlic, chives, coriander, or anise
Bean leaf beetle	Potato, onion, or turnip
Codling moth	Common oleander
Colorado potato bug	Green beans, coriander, or nasturtium
Cucumber beetle	Radish or tansy
Flea beetle	Garlic, onion, or mint
Imported cabbage worm	Mint, sage, rosemary, or hyssop
Japanese beetle	Garlic, larkspur, tansy, rue, or geranium
Leaf hopper	Geranium or petunia
Mice	Onion
Root knot nematodes	French marigolds
Slugs	Prostrate rosemary or wormwood
Spider mites	Onion, garlic, cloves, or chives
Squash bug	Radish, marigolds, tansy, or nasturtium
Stink bug	Radish
Thrips	Marigolds
Tomato hornworm	Marigolds, sage, or borage
Whitefly	Marigolds or nasturtium

also mimic some disease problems. Yellowing or reddening of leaves and stunted growth may indicate a nutritional problem. Leaf curling or misshapen growth may be a result of herbicide application.

Pest and Disease Management Practices

Preventing pests should be your first goal when growing a garden, although it is unlikely that you will be able to avoid all pest problems because some plant seeds and disease organisms may lay dormant in the soil for years.

Diseases need three elements to become established in plants: the disease organism, a susceptible species, and the proper environmental conditions. Some disease organisms can live in the soil for years; other organisms are carried in infected plant material that falls to the ground. Some disease organisms are carried by insects. Good sanitation will help limit some problems with disease. Choosing resistant varieties of plants also prevents many diseases from occurring. Rotating annual plants in a garden can also prevent some diseases.

Plants that have adequate, but not excessive, nutrients are better able to resist attacks from both diseases and insects. Excessive rates of nitrogen often result in extremely succulent vegetative growth and can make plants more susceptible to insect and disease problems, as well as decreasing their winter hardiness. Proper watering and spacing of plants limits the spread of some diseases and provides good aeration around plants, so diseases that fester in standing water cannot multiply. Trickle irrigation, where water is

applied to the soil and not the plant leaves, may be helpful.

Removal of diseased material certainly limits the spread of some diseases. It is important to clean up litter dropped from diseased plants. Prune diseased branches on trees and shrubs to allow for more air circulation. When pruning diseased trees and shrubs, disinfect your pruners between cuts with a solution of chlorine bleach to avoid spreading the disease from plant to plant. Also try to control insects that may carry diseases to your plants.

You can make your own natural fungicide by combining 5 teaspoons each of baking soda and hydrogen peroxide with a gallon of water. Spray on your infected plants. Milk diluted with water is also an effective fungicide, due to the potassium phosphate in it, which boosts a plant's immune system. The more diluted the solution, the more frequently you'll need to spray the plant.

« Powdery mildew leaf disease

Harvesting Your Garden

It is essential, in order to get the best freshness, flavor, and nutritional benefits from your garden vegetables and fruits, to harvest them at the appropriate time. The vegetable's stage of maturity and the time of day at which it is harvested are essential for good-tasting and nutritious produce. Overripe vegetables and fruits will be stringy and coarse. When possible, harvest your vegetables during the cool part of the morning. If you are going to can and preserve your vegetables and fruits, do so as soon as possible. Or, if this process must be delayed, make sure to cool the vegetables in ice water or crushed ice and store them in the refrigerator. Here are some brief guidelines for harvesting various types of common garden produce:

Asparagus—Harvest the spears when they are at least 6 to 8 inches tall by snapping or cutting them at ground level. A few spears may be harvested the second year after crowns are set out. A full harvest season will last four to six weeks during the third growing season.

Beans, snap—Harvest before the seeds develop in the pod. Beans are ready to pick if they snap easily when bent in half.

Beans, lima—Harvest when the pods first start to bulge with the enlarged seeds. Pods must still be green, not yellowish.

Broccoli—Harvest the dark green, compact cluster, or head, while the buds are shut tight, before any yellow flowers appear. Smaller side shoots will develop later, providing a continuous harvest.

Brussels sprouts—Harvest the lower sprouts (small heads) when they are about 1 to 1½ inches in diameter by twisting them off. Removing the lower leaves along the stem will help to hasten the plant's maturity.

Cabbage—Harvest when the heads feel hard and solid.

Cantaloupe—Harvest when the stem slips easily from the fruit with a gentle tug. Another indicator of ripeness is when the netting on the skin becomes rounded and the flesh between the netting turns from a green to a tan color.

Carrots—Harvest when the roots are ¾ to 1 inch in diameter. The largest roots generally have darker tops.

» Only the eggplant vegetable is edible. Do not eat the leaves, stem, roots, or flowers.

Cauliflower—When preparing to harvest, exclude sunlight when the curds (heads) are 1 to 2 inches in diameter by loosely tying the outer leaves together above the curd with a string or rubber band. This process is known as blanching. Harvest the curds when they are 4 to 6 inches in diameter but still compact, white, and smooth. The head should be ready ten to fifteen days after tying the leaves.

Collards—Harvest older, lower leaves when they reach a length of 8 to 12 inches. New leaves will grow as long as the central growing point remains, providing a continuous harvest. Whole plants may be harvested and cooked if desired.

Corn, sweet—The silks begin to turn brown and dry out as the ears mature. Check a few ears for maturity by opening the top of the ear and pressing a few kernels with your thumbnail. If the exuded liquid is milky rather than clear, the ear is ready for harvesting. Cooking a few ears is also a good way to test for maturity.

Cucumbers—Harvest when the fruits are 6 to 8 inches in length. Harvest when the color is deep green and before yellow color appears. Pick four to five times per week to encourage continuous production. Leaving mature cucumbers on the vine will stop the production of the entire plant.

Eggplant—Harvest when the fruits are 4 to 5 inches in diameter and their color is a glossy, purplish black. The fruit is getting too ripe when the color starts to dull or become bronzed. Because the stem is woody, cut—do not pull—the fruit from the plant. A short stem should remain on each fruit.

Kale—Harvest by twisting off the outer, older leaves when they reach a length of 8 to 10 inches and are medium green in color. Heavy, dark green leaves are overripe and are likely to be tough and bitter. New leaves will grow, providing a continuous harvest.

Lettuce—Harvest the older, outer leaves from leaf lettuce as soon as they are 4 to 6 inches long. Harvest heading types when

the heads are moderately firm and before seed stalks form.

Mustard—Harvest the leaves and leaf stems when they are 6 to 8 inches long; new leaves will provide a continuous harvest until they become too strong in flavor and tough in texture due to temperature extremes.

Okra—Harvest young, tender pods when they are 2 to 3 inches long. Pick the okra at least every other day during the peak growing season. Overripe pods become woody and are too tough to eat.

Onions—Harvest when the tops fall over and begin to turn yellow. Dig up the onions and allow them to dry out in the open sun for a few days to toughen the skin. Then remove the dried soil by brushing the onions lightly. Cut the stem, leaving 2 to 3 inches attached, and store in a net-type bag in a cool, dry place.

Peas—Harvest regular peas when the pods are well rounded; edible-pod varieties should be harvested when the seeds are fully developed but still fresh and bright green. Pods are getting too old when they lose their brightness and turn light or yellowish green.

Peppers—Harvest sweet peppers with a sharp knife when the fruits are firm, crisp, and full size. Green peppers will turn red if left on the plant. Allow hot peppers to attain their bright red color and full flavor while attached to the vine; then cut them and hang them to dry.

Potatoes (Irish)—Harvest the tubers when the plants begin to dry and die down. Store the tubers in a cool, high-humidity location with good ventilation, such as the basement or crawl space of your house. Avoid exposing the tubers to light, as greening, which denotes the presence of dangerous alkaloids, will occur even with small amounts of light.

Pumpkins—Harvest pumpkins and winter squash before the first frost. After the vines dry up, the fruit color darkens and the skin surface resists puncture from your thumbnail. Avoid bruising or scratching the fruit while handling it. Leave a 3- to 4-inch portion of the stem attached to the fruit and store it in a cool, dry location with good ventilation.

Radishes—Harvest when the roots are ½ to 1½ inches in diameter. The shoulders of radish roots often appear through the soil surface when they are mature. If left in the ground too long, the radishes will become tough and woody.

Rutabagas—Harvest when the roots are about 3 inches in diameter. The roots may be stored in the ground and used as needed, if properly mulched.

Spinach—Harvest by cutting all the leaves off at the base of the plant when they are 4 to 6 inches long. New leaves will grow, providing additional harvests.

Squash, summer—Harvest when the fruit is soft, tender, and 6 to 8 inches long. The skin color often changes to a dark, glossy green or yellow, depending on the variety. Pick every two to three days to encourage continued production.

Sweet potatoes—Harvest the roots when they are large enough for use before the first frost. Avoid bruising or scratching the potatoes during handling. Ideal storage conditions are at a temperature of 55 degrees Fahrenheit and a relative humidity of 85 percent. The basement or crawl space of a house may suffice.

Swiss chard—Harvest by breaking off the developed outer leaves 1 inch above the soil. New leaves will grow, providing a continuous harvest.

Tomatoes—Harvest the fruits at the most appealing stage of ripeness, when they are bright red. The flavor is best at room temperature, but ripe fruit may be held in the refrigerator at 45 to 50 degrees Fahrenheit for seven to ten days.

Turnips—Harvest the roots when they are 2 to 3 inches in diameter but before heavy fall frosts occur. The tops may be used as salad greens when the leaves are 3 to 5 inches long. Turnips can be eaten almost any way potatoes can be—mashed, roasted, or even fried. The greens can be eaten raw, steamed, or boiled.

Watermelons—Harvest when the watermelon produces a dull thud rather than a sharp, metallic sound when thumped—this means the fruit is ripe. Other ripeness indicators are a deep yellow rather than a white color where the melon touches the ground, brown tendrils on the stem near the fruit, and a rough, slightly ridged feel to the skin surface.

Part 2 The Country Kitchen

Baking Bread 52
Maple Sugaring 61
Making Sausage 64
The Home Dairy 72
Canning 82
Drying and Freezing 155

The kitchen is a country in which there are always discoveries to be made.

—*Grimod de la Reyniere*, 1758–1838

Baking Bread

Bread has been a dining staple for thousands of years. The art of bread making has evolved over time, but the basic principles remain unchanged. Bread is made from flour of wheat or other grains, with the addition of water, salt, and a fermenting ingredient (such as yeast or another leavening agent). After you've baked a few loaves, you'll start to get a feel for what the dough should look and feel like. Then you can start experimenting with different flours, or additions of fruits, nuts, seeds, herbs, and more.

Quick Breads

Muffins, banana bread, zucchini bread, and many other sweet breads are often leavened with agents other than yeast, such as baking soda or baking powder. These breads are easy to make and require far less preparation time than yeast breads. They're also very versatile; once you master the basic recipe you can add almost any fruit, nut, or flavoring to make a uniquely delicious treat.

Basic Quick Bread Recipe

This basic recipe will make two loaves or twelve large muffins. Fold in 1 to 2 cups of mashed fruit, whole berries, nuts, or chocolate chips before pouring the batter into the pans.

- 3½ cups flour (use at least 2 cups of a gluten-rich flour)
- 2 teaspoons baking powder
- 1 teaspoon baking soda
- 1 teaspoon salt
- 1 to 2 teaspoons spices or herbs, if desired
- 1¼ cups sugar
- ¾ cup butter, oil, or fruit puree
- 3 eggs
- ¾ cup milk

1. In a large mixing bowl combine all dry ingredients except sugar.
2. In a separate bowl, beat together sugar and butter, oil, or fruit puree. Add eggs and beat until light and fluffy.
3. Add butter and sugar mixture and milk alternately to the dry ingredients, stirring just until combined. Fold in additional fruit, nuts, or flavors of your choice.
4. For bread, pour into a greased bread pan and bake at 350°F for 1 hour. For muffins, fill muffin cups 2/3 full and bake at 350°F for 20 to 25 minutes.

Cinnamon Bread

- 2 eggs
- ½ cup butter
- 1 cup sugar
- ½ cup milk
- 1¼ cups flour
- 2½ teaspoons baking powder
- 1 teaspoon cinnamon
- 1 teaspoon butter, melted
- 2 tablespoons sugar and 2 tablespoons cinnamon, mixed together

1. Beat together the eggs, butter, and sugar until fluffy.
2. In a separate bowl, combine the dry ingredients. Add the dry mixture and the milk to the butter mixture and mix until combined.
3. Bake in a greased bread pan at 300°F for almost an hour. When done, pour melted butter over top and sprinkle with cinnamon and sugar mixture.

TIP

Baking powder is a mixture of baking soda, cornstarch, and cream of tartar in a 1:1:2 ratio. To make 1 teaspoon of baking powder, combine ¼ teaspoon baking soda, ¼ teaspoon cornstarch, and ½ teaspoon cream of tartar.

One-Hour Brown Bread

- 1 cup cornmeal
- 1 cup white flour
- ½ teaspoon salt
- 1 teaspoon baking soda
- 1 cup water, boiling
- 1 egg
- ½ cup molasses
- ½ cup sugar

1. Combine cornmeal, flour, and salt.
2. Add the baking soda to boiling water and stir. Add to dry ingredients.
3. Beat together egg, molasses, and sugar and add to dry ingredients. Mix until combined. Pour batter into an empty coffee can with a cover (or cover with foil).
4. Place a cake rack in the bottom of a dutch oven or large pot. Place the covered can on the rack and pour boiling water into the pot until it reaches half way up the can. Cover the pot, turn the unit on very low, and steam for one hour.

Cranberry Coffee Cake

- 2 tablespoons butter
- ¼ cup firmly packed brown sugar
- 1 cup cooked or canned cranberry sauce
- ¼ cup pecans, chopped

- 1 tablespoon grated orange rind
- 1½ cups sifted flour
- 2 teaspoons, double acting baking powder
- ¼ cup sugar
- ⅓ cup shortening
- 1 egg, beaten
- ½ cup milk

1. Melt butter in 9-inch ring mold. Spread brown sugar over bottom of pan.
2. Combine cranberry sauce, pecans, and orange rind. Spread over brown sugar in bottom of pan.
3. Sift together flour, baking powder, and sugar.
4. Cut in shortening until dough resembles coarse meal. Combine egg with milk. Add all at once, mixing only to dampen flour. Turn into pan.
5. Bake at 400°F for 25 to 30 minutes. Cool 5 minutes and invert onto plate. Serve warm.

Date-Orange Bread

- 2 tablespoons butter or margarine, melted
- ¾ cup orange juice
- 2 tablespoons grated orange rind
- ½ cup finely cut dates
- 1 cup sugar
- 1 egg, slightly beaten
- ½ cup coarsely chopped pecan
- 2 cups sifted all-purpose flour
- ½ teaspoon baking soda
- 1 teaspoon baking powder
- ½ teaspoon salt

1. Combine first seven ingredients.
2. Mix and sift remaining ingredients; stir into wet mixture. Mix well, but quickly, being careful not to overbeat.
3. Turn into greased loaf pan. Bake in moderate oven, 350°F, for 50 minutes or until done. Remove from pan and let cool right side up, on a wire rack.

Pineapple Nut Bread

- 2¼ cups sifted flour
- ¾ cup sugar
- 1½ teaspoons salt
- 3 teaspoons baking powder
- ½ teaspoon baking soda
- 1 cup prepared bran cereal
- ¾ cup walnuts, chopped
- 1½ cups crushed pineapple, undrained
- 1 egg, beaten
- 3 tablespoons shortening, melted

1. Sift flour, sugar, salt, baking powder, and soda together.
2. Mix together remaining ingredients and combine with dry mixture.
3. Bake in greased loaf pan at 350°F for 1 ¼ hrs. This bread keeps moist a week or ten days, and slices best when a day or more old.

Date Muffins

- 1¾ cups sifted enriched flour
- 2 tablespoons sugar
- 2¼ teaspoons baking powder
- ¾ teaspoon salt
- ½ to ¾ cup coarsely cut pitted dates
- 1 egg, well-beaten
- ¾ cup milk
- ⅓ cup melted shortening or salad oil

1. Sift dry ingredients into mixing bowl and stir in dates. Make a well in center.
2. Combine egg, milk, and salad oil; add all at once to dry ingredients. Stir quickly only till dry ingredients are moistened.
3. Drop batter by tablespoons into greased muffin pans. Fill two-thirds full. Bake in oven, at 400°F, for about 25 minutes. Makes one dozen.

Yeast Bread

Once you've made a loaf of homemade yeast bread, you'll never want to go back to buying packaged bread from the grocery store. Homemade bread tastes and smells heavenly and the baking process itself can be very rewarding. Store homemade bread in a paper or resealable plastic bag and eat within a day or two for best results. Bread that begins to get stale can be cubed and made into stuffing or croutons.

Before you start baking, it's helpful to understand the various components that make up bread.

Wheat

Wheat is the most common flour used in bread making, as it contains gluten in the right proportion to make bread rise. Gluten, the protein of wheat, is a gray, tough, elastic substance, insoluble in water. It holds the gas developed in bread dough by fermentation, which otherwise would escape. Though there are many ways to make gluten-free bread, flour that naturally contains gluten will rise more easily than gluten-free grains. In general, combining smaller amounts of other flours (rye, corn, oat, etc.) with a larger proportion of wheat flour will yield the best results.

A grain of wheat consists of (1) an outer covering, or husk, which is always removed before milling; (2) bran, a hard shell that contains minerals and is high in fiber; (3) the germ, which contains the fat and protein content and is the part that can be planted and cultivated to grow more wheat; and (4) the endosperm, which is the wheat plant's own food source and is mostly starch and protein. Whole wheat contains all of these components except for the husk. White flour is only the endosperm.

Yeast

Yeast is a microscopic fungus that consists of spores, or germs. These spores grow by budding and division, multiply very rapidly under favorable conditions, and produce fermentation. Fermentation is the process by which, under influence of air, warmth, and a fermenting ingredient, sugar (or dextrose, starch converted into sugar) is changed into alcohol and carbon dioxide.

Dry yeast is most commonly used for baking. Most grocery stores sell regular active dry and instant yeast. Instant yeast is more finely ground and thus absorbs the moisture faster, speeding up the leavening process and making the bread rise more rapidly.

Active dry yeast should be proofed before using. Mix one packet of active dry yeast with ¼ cup warm water and 1 teaspoon sugar. Stir until yeast dissolves. Allow it to sit for 5 minutes, or until it becomes foamy.

Making Bread

Making bread is a fairly simple process, though it does require a chunk of time. Keep in mind, though, that you can be doing other things while the bread is rising or baking. The process is fairly straightforward and only varies slightly by kind of bread.

Mix together the flour, sugar, salt, and any other dry ingredients. Form a well in the center and add the dissolved yeast and any other wet ingredients. Mix all the ingredients together.

Gather the dough into a ball and place it on a lightly floured surface. Flour your hands to keep the dough from sticking to your fingers. Knead the dough by folding it toward you and then pushing it away with the palms of your hands. Continue kneading for five to ten minutes, or until the dough is soft and elastic.

Place the dough in a lightly greased pan, cover with a dish towel, and allow to rise in a warm place until it doubles in size.

Punch the dough down to expel the air and place it in a greased and lightly floured baking pan. Cover and let rise a second time until it doubles in size.

Bake the bread in a preheated oven according to the recipe. Bread is done when it is golden brown and sounds hollow when you tap the top.

Remove bread from the pan by loosening the sides with a knife or spatula and tipping the pan upside down onto a wire rack.

Biscuits

Any bread recipe can be made into biscuits instead of one large loaf. To shape bread dough into biscuits, pull or cut off pieces, making them all as close to uniform in size as possible. Flour palms of hands slightly and shape each piece individually. Using the thumb and first two fingers of one hand, and holding it in the palm of the other hand, move the dough round and round, folding the dough towards the center. When smooth, turn it over and roll between palms of hands. Place in greased pans nearly together, brushed between with a little melted butter, which will allow biscuits to separate after baking.

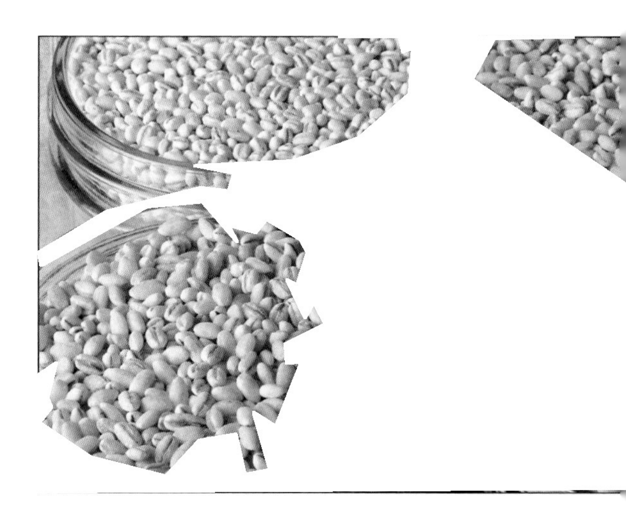

FLOUR	DESCRIPTION
All-purpose	A blend of high- and low-gluten wheat. Slightly less protein than bread flour. Best for cookies and cakes.
Amaranth	Gluten-free. Made from seeds of amaranth plants. Very high in fiber and iron.
Arrowroot	Gluten-free. Made from the ground-up root. Clear when cooked, which makes it perfect for thickening soups or sauces.
Barley	Ground barley grain. Very low in gluten. Used as a thickener in soups or stews or mixed with other flours in baked goods.
Bran	Made from the hard outer layer of wheat berries. Very high in protein, fiber, vitamins, and minerals.
Bread	Made from hard, high-protein wheat with small amounts of malted barley flour and vitamin C or potassium bromate. It has a high gluten content, which helps bread to rise. Excellent for bread, but not as good for use in cookies or cakes.
Buckwheat	Gluten-free. Highly nutritious with a slightly nutty flavor.
Chickpea	Gluten-free. Made from ground chickpeas. Used frequently in Indian, Middle Eastern, and some French Provençal cooking.
Oat	Gluten-free. Made from ground oats. High in fiber.
Quinoa	Gluten-free. Made from ground quinoa, a grain native to the Andes in South America. Slightly yellow or ivory-colored with a mild nutty flavor. Very high in protein.
Rye	Milled from rye berries and rye grass. High in fiber and low in gluten. Light rye has had more of the bran removed through the milling process than dark rye. Slightly sour flavor.
Semolina	Finely ground endosperm of durum wheat. Very high in gluten. Often used in pasta.
Soy	Gluten-free. Made from ground soybeans. High in protein and fiber.
Spelt	Similar to wheat, but with a higher protein and nutrient content. Contains gluten but is often easier to digest than wheat. Slightly nutty flavor.
Tapioca	Gluten-free. Made from the cassava plant. Starchy and slightly sweet. Generally used for thickening soups or puddings, but can also be used along with other flours in baked goods.
Teff	Gluten-free. Higher protein content than wheat and full of fiber, iron, calcium, and thiamin.
Whole wheat	Includes the bran, germ, and endosperm of the wheat berry. Far more nutritious than white flour, but has a shorter shelf life.

MILLING YOUR OWN GRAINS

You can grind grains into flour at home using a mortar and pestle, a coffee or spice mill, manual or electric food grinders, a blender, or a food processor. Grains with a shell (quinoa, wheat berries, etc.) should be rinsed and dried before milling to remove the layer of resin from the outer shell that can impart a bitter taste to your flour. Rinse the grains thoroughly in a colander or mesh strainer, then spread them on a paper or cloth towel to absorb the extra moisture. Transfer to a baking sheet and allow to air dry completely (to speed this process, you can put them in a very low oven for a few minutes). When the grains are dry, they're ready to be ground.

Multigrain Bread

- ¼ cup yellow cornmeal
- ¼ cup packed brown sugar
- 1 teaspoon salt
- 2 tablespoons vegetable oil
- 1 cup boiling water
- 1 package active dry yeast
- ¼ cup warm (105 to 115°F) water
- ¼ cup whole wheat flour
- ¼ cup rye flour
- 2¼–2¾ cups all-purpose flour

1. Mix cornmeal, brown sugar, salt, and oil with boiling water; cool to lukewarm (105 to 115°F).
2. Dissolve yeast in ¼ cup warm water; stir into cornmeal mixture. Add whole wheat and rye flours and mix well. Stir in enough all-purpose flour to make dough stiff enough to knead.
3. Turn dough onto lightly floured surface. Knead until smooth and elastic, about 5 to 10 minutes.
4. Place dough in lightly oiled bowl, turning to oil top. Cover with clean towel; let rise in warm place until double, about 1 hour.
5. Punch dough down; turn onto clean surface. Cover with clean towel; let rest 10 minutes. Shape dough and place in greased 9 x 5 inch pan. Cover with clean towel; let rise until almost double, about 1 hour.
6. Preheat oven to 375°F. Bake 35 to 45 minutes or until bread sounds hollow when tapped. Cover with aluminum foil during baking if bread is browning too quickly. Remove bread from pan and cool on wire rack.

TIP

- 1 package active dry yeast = about 2 ¼ teaspoons = ¼ ounce
- 4-ounce jar active dry yeast = 14 tablespoons
- 1 (6-ounce) cube or cake of compressed yeast (also known as fresh yeast) = 1 package of active dry yeast
- Multiply the amount of instant yeast by 3 for the equivalent amount of fresh yeast
- Multiply the amount of active dry yeast by 2.5 for the equivalent amount of fresh yeast
- Multiply the amount of instant yeast by 1.25 for the equivalent amount of active dry yeast

Oatmeal Bread

- 1 cup rolled oats
- 1 teaspoon salt
- 1½ cups boiling water
- 1 package active dry yeast
- ¼ cup warm water (105 to 115°F)
- ¼ cup light molasses
- 1½ tablespoons vegetable oil
- 2 cups whole wheat flour
- 2–2½ cups all-purpose flour

1. Combine rolled oats and salt in a large mixing bowl. Stir in boiling water; cool to lukewarm (105 to 115°F).
2. Dissolve yeast in ¼ cup warm water in small bowl.
3. Add yeast water, molasses, and oil to cooled oatmeal mixture. Stir in whole wheat flour and 1 cup all-purpose flour. Add additional all-purpose flour to make a dough stiff enough to knead.
4. Knead dough on lightly floured surface until smooth and elastic, about 5 minutes.
5. Place dough in lightly oiled bowl, turning to oil top. Cover with clean towel; let rise in warm place until double, about 1 hour.
6. Punch dough down; turn onto clean surface. Shape dough and place in greased 9 x 5 inch pan. Cover with clean towel; let rise in a warm place until almost double, about 1 hour.
7. Preheat oven to 375°F. Bake 50 minutes or until bread sounds hollow when tapped. Cover with aluminum foil during baking if bread is browning too quickly. Remove bread from pan and cool on wire rack.

Raised Buns (Brioche)

- ½ cup milk
- ⅓ cup butter
- ¼ cup sugar
- ¾ teaspoon salt
- 1 package yeast (active, dry, or compressed)
- ¼ cup lukewarm water
- 3 eggs, well-beaten
- 2½ cups enriched flour
- ¼ teaspoon lemon rind, or ⅛ teaspoon crushed cardamom seeds

GLUTEN-FREE BREAD

Making good gluten-free bread isn't always easy, but there are several things you can do to improve your chances of success:

- Choose flours that are high in protein, such as sorghum, amaranth, millet, teff, gluten-free oatmeal, and buckwheat
- Use all room temperature ingredients. Yeast thrives in warm environments.
- Add a couple teaspoons of xantham gum to your dry ingredients.
- Add eggs and dry milk powder to your bread. These will add texture and help the bread to rise.
- Crush a vitamin C tablet and add it to your dry ingredients. The acidity will help the yeast do its job.
- Substitiute carbonated water or gluten-free beer for other liquids in the recipe.
- If you're following a traditional bread recipe, add extra liquid (water, carbonated water, milk, fruit juice, or olive oil) to get a soft and sticky consistency. The batter should be a little too sticky to knead. For this reason, bread machines are great for making gluten-free bread.

1. Scald milk; stir in butter, sugar, and salt; cool to lukewarm. Sprinkle or crumble yeast into water in large bowl; stir to dissolve. Add lukewarm milk mixture and beaten eggs, setting aside 2 tablespoons beaten egg to brush on the brioches before baking.; mix well.
2. Sift flour; add 1½ cups of it to mixture and beat by hand 8 minutes or with electric mixer at medium speed for 3 minutes. Add remaining sifted flour and lemon rind; beat to smooth, heavy batter. Cover with towel; let rise in warm place 2 hours or until doubled. Stir down; cover tightly; chill at least 5 hours or overnight.
3. Stir down again. Mixture is soft now. Grease hands slightly, place dough on lightly floured board, and knead a few times. With sharp knife, cut off small pieces of dough. Roll with hands to about ½ inch in diameter, and about 10 inches long. Coil loosely in circle, winding around toward center. Top with small ball of dough. Cover with a towel and let rise ½ hour or until doubled. Brush with leftover beaten egg. Bake in oven 400°F for 12 to 15 minutes. For an extra touch, top with thin frosting while still warm. Makes 12 to 18 buns.

Maple Sugaring

The production of maple syrup and maple sugar is purely an American industry, Canada being the only country outside of the United States where they are made. The earliest explorers in this country found the Native Americans making sugar from maple trees, and in some sections producing it in quantity for trade. The settlers began to make maple products as well and to attempt to improve their manufacture. For many years, maple sugar was the only sugar used and, despite refinements, beyond the tapping and boiling, the general process remains the same as at that time.

All the maples have sweet sap, but only from a few of the species has sugar been made in worthwhile quantities. The first place is held by the sugar maple and a variety of it—the black maple. These can be found in the Northeastern region of the United States, as well as the northern Midwest. Other varieties, including the red maple, the silver maple, and the Oregon maple, can be tapped, but will produce smaller quantities of syrup. It takes approximately 40 gallons of sap to make 1 gallon of syrup, so it is generally not worthwhile to tap the less productive tree varieties.

Tapping

The quantity of sap that a tree yields stands in direct relation to the size of its crown. It is good to make it a rule to

tap only one place on a tree; by doing so the life of the tree is prolonged. Large trees might be tapped in two and sometimes three places without injury, but not in two places so near together that the sap from the two is collected in one bucket. Each hole should heal over in as quickly as one season.

Before tapping, the side of the tree should be brushed with a stiff broom to remove all loose bark and dirt and a spot selected where the bark looks healthy, some distance from the scar of a previous tapping. Care should also be taken to tap where a bucket attached to the spout inserted in the hole will hang level and be partly supported by the tree itself. The distance from the ground should be about waist high, convenient for the sap collector. In general, it is best to tap on the side of the tree where other trees do not shade the spot. The main requisite in tapping a tree is a good sharp bit with which a clean-cut hole can be made. A rough, feathered hole soon becomes foul, stopping the flow. After the tapping, all shavings should be removed to make the hole clean. The bark should never be cut away before boring the hole, as this shortens the life of the tree.

General practice concerning the size of the hole seems to indicate that three-eighths to half an inch is the best diameter; then, if the season is long and a warm spell interrupts the flow, the holes can be reamed out to ½ to ⅝ of an inch, and thereby secure an increased run. A 13/32 of an inch bit is often used. The bit should be especially sharp and should bring the shavings to the surface. Its direction is slightly upward into the tree. The slant allows the hole to drain readily.

The depth of the hole should be regulated by the size of the tree, as only the layers next to the bark are alive and contain enough sap to flow freely. Toward the interior, the flow diminishes. With the ordinary tree a hole less than 1½ to 2 inches deep is best. In small trees, make a short incision just through the sapwood. In any case, boring should be stopped when dark colored shavings appear, as this shows dead wood and that the sapwood has been passed through. It is good policy to tap early in the season in order to obtain the earlier runs, which are generally the sweetest. "Sugar weather" begins sometime between mid-February to mid-March, when the days are becoming warm, the temperature going above 32°F, and the nights are still frosty.

The spout, or spile, is the tube through which the sap flows into the bucket. It is usually of metal, but hollow reeds are sometimes used. The best are perfectly cylindrical and of an even taper, making them easy to insert and to remove without interfering with the wood tissue. The perfect spout should be strong enough to support the bucket of sap safely, and for obvious reasons should bring the whole weight on the bark of the tree and not on the inner tissue or sapwood. A spout should have a hook or stop on which the bucket is to hang, unless the bucket may hang on the spout itself, and it is best to have a spout with a small hole, because one with a large hole allows the bore to dry out faster when there are strong winds. Buckets are typically of galvanized metal (free from corrosion or rust, covered, and fitting well to the tree). Sometimes old plastic one-gallon milk cartons are used, since their narrow neck prevents debris from entering. Make sure to clean any container thoroughly before use.

The sap should be collected each day and not be allowed to accumulate. It is also necessary to keep the buckets and containers clean, and they should be washed in warm water after each run. So long as it is cold, you may store the sap outdoors for up to three days in any large metal or plastic container. When pouring the sap into its collecting device, stretch a flannel cloth over the top of the tanks and pour the sap through this to remove any twigs, leaves, or pieces of dirt.

Syrup and Sugar Making

Once you have enough sap to start making syrup, you may start to boil it down. Use any outdoor method, from bonfire to coal-burning range, camp stove to commercial evaporator, but avoid boiling sap inside, as it results in a sticky residue on your walls.

Use two pans, one to evaporate excess moisture from the sap and concentrate it into syrup, and one as a finishing pan, in which you will finish boiling it. The evaporator should have a large bottom surface area, and the sap should not be deeper than 1 to 1½ inches in the pan at any time. The size of the pan depends on how much sugaring you intend to do—it is best if it can hold at least one gallon. Put in an inch or two of sap, boil, and add more, a little at a time, so as not to stop boiling or materially change the density of the boiling liquid; then, when this charge is concentrated, or has reached approximately 6 degrees above the temperature at which water boils (use a candy thermometer to monitor it), the syrup should be drawn off. Care must be exercised not to allow the remaining syrup in the pan to be burned. While evaporating, use a kitchen strainer to skim off the froth, and keep a spoon for stirring on hand.

To finish the syrup, pour it through a piece of felt cloth, or two pieces of thick flannel, into the finishing pot. Once the syrup has reached 7 degrees above the boiling point of water, it is ready for storage.

Glass containers are the best method for keeping syrup, although airtight plastic and metal containers will also work. When carefully canned, syrup will keep from one season to another without souring or bursting the jar. It is best to store syrup immediately after finishing, while still hot, and then keep at an even, cold temperature. Temperatures around freezing, however, should not be used, as this may crystallize the syrup.

"Sugaring-off" applies to the further treatment of the maple syrup by which it is made into a solid product. The ordinary iron pot of the kitchen is filled nearly half full with the syrup and this is concentrated over the fire. Use a candy thermometer to determine the proper point of stopping the boiling. In the first runs of sap, the boiling should be carried up to 26 to 28 degrees above the boiling point of water at that elevation to make a medium hard sugar. With later runs, the finishing temperature should be 28 to 38 degrees above the boiling point. After the thick syrup has reached the proper boiling point, it should be taken from the fire and stirred until somewhat cooled. This gives it a uniform grain and color in the mold. As in syrup making, one should "sugar-off" a charge before adding any more syrup. The hardness of the sugar produced is to a large extent controlled by its moisture content. High temperatures are required to evaporate more of the water, but note that for softer sugars, you should use slightly lower temperatures.

Like brown sugar, maple sugar does not keep well in a moist atmosphere. It tends to absorb water, molds rather quickly, and if finished at too low a temperature, the sugar is soft and the liquid portion drains out. Therefore sugar which is to be stored should always be boiled to a high temperature. It can be wrapped in paper, but should not be put in covered containers unless these are absolutely sealed. It is best to store the sugar in a warm room of even temperature. If the cakes are sealed without access to air, a cold place can be used, but make sure they are kept dry.

Making Sausage

Important Considerations in Sausage Making

Temperatures

Meat products are extremely perishable and must be maintained under refrigeration (40°F or below). When you have finished processing a product, return it to the refrigerator. After the product has been formulated, smoke and cook the product to the required temperature and then return the product to refrigeration. Sausage that hasn't been properly refrigerated will spoil and can make you sick.

Sanitation

There is no substitute for keeping the tables, utensils, and ingredients clean and free from dirt and contamination. Use plenty of hot water and soap before and after processing sausages. Always keep your hands clean.

Taking Notes

Just as you keep a copy of a good recipe, you should keep notes on the formulation and processing procedures of your favorite smoked and cooked sausage. Ingredients, times, temperatures, and end results should be noted. This will help to make a better sausage the next time.

Fat Content

Different sausages have different amounts of fat. Avoid making the formula too lean as the sausage will be dry and hard. Fresh pork sausage contains 30 to 45 percent fat. Smoked or roasted sausage contains 20 to 30 percent fat. Formulate the fat content just as you would the other ingredients in a sausage.

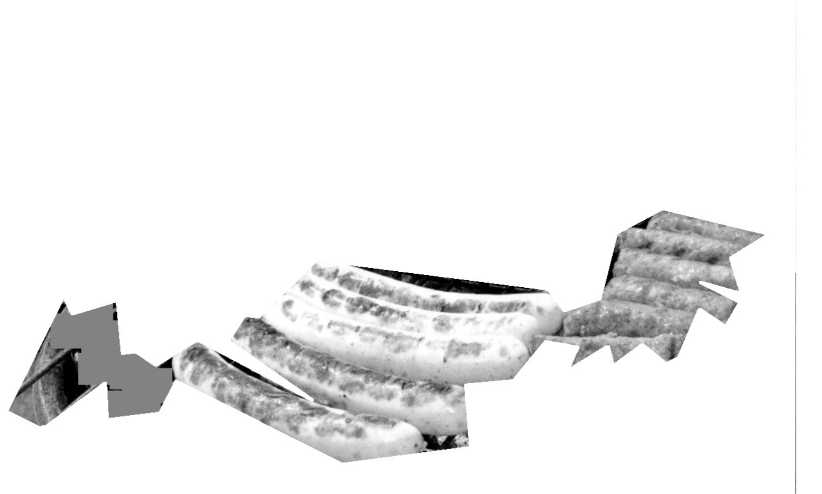

Nitrates and Nitrites

These curing ingredients are required to achieve the

characteristic flavor, color, and stability of cured meat. Nitrate and nitrite are converted to nitric oxide by microorganisms and combine with the meat pigment myoglobin to give the cured meat color. However, more importantly, nitrite provides protection against the growth of botulism-producing organisms, acts to show rancidity, and stabilizes the flavor of the cured meat. Potassium nitrate (saltpeter) was the salt historically used for curing. However, we now know that saltpeter is much stronger than necessary for curing meat.

Much controversy has surrounded the use of nitrite in recent years—for good reason. Too many nitrites can cause all sorts of health problems, including cancer. However, the amount used in sausage making is not enough to cause concern and is far less of a worry than the botulism that can result if it is not used. A commercial premixed cure can be used when nitrate or nitrite is called for in the recipe. The premixes have been diluted with salt so that the small quantities of nitrites and nitrates can more easily be weighed. This reduces the possibility of serious error in handling pure nitrate or nitrite. Several premixes are available. Many local grocery stores stock Morton® Tender Quick® Product and other brands of premix cure.

Because the amount of premixed cure will vary depending on what brand is used, it is important to follow the directions on the package. The recipes below are only for fresh sausage, which does not require a cure. Fresh sausage is delicious, simple to make, and can be frozen if you do not want to cook and serve it right away.

Storage

The length of time a sausage can be stored depends on the type of sausage. Fresh sausage is highly perishable and will only last seven to ten days. However, it may be frozen for four to six months if wrapped in moisture-vapor-proof wrap (freezer paper). Smoked sausages may last from two to four weeks under refrigeration.

Fresh Pork Sausage

Fresh pork sausage is a mixture of pork meats, salt, and spices which have been ground or chopped with no added water or extenders. Fat content usually ranges from 35 to 50 percent depending upon individual preference. For a spicier sausage, add a teaspoon of crushed red pepper or ½ teaspoon ground red pepper.

Ingredients

- 1½ lbs. 60% lean ground pork
- 1 teaspoon sugar
- 1 teaspoon salt
- 3 tablespoons fresh minced parsley
- 1 teaspoon sage

Directions

Mix spices with ground meat. Form into nine small patties and cook in a cast iron skillet until brown on both sides and cooked in the middle. Stuff in natural casings (pork rounds) or collagen casings.

All Beef Sausage

This recipe can be halved or quartered for smaller portions. You can substitute 85% lean ground beef for the beef chucks and plates in this recipe if you don't want to pull out the meat grinder. You can also mix up a large batch of the spices ahead of time to have on hand and use a portion every time you make sausage.

Ingredients

- 7½ lbs. fresh boneless beef chucks (85% lean)
- 2½ lbs. fresh boneless beef plates (50% fat)
- ½ cup salt
- 2 tablespoons sugar
- 3 tablespoons ground white pepper
- 1 teaspoon mashed peeled garlic
- 1½ tablespoons paprika
- 2½ teaspoons ground coriander

1. Grind beef chucks through ⅜-inch plate and beef plates through ¼-inch plate. Combine meat ingredients adding salt, sugar, and seasoning and mix well for five minutes.
2. Transfer to stuffer and stuff in cellulose or fibrous casings three to four inches in diameter or large beef casing. Refrigerate or freeze immediately.

Polish Sausage *(Kielbasa)*

Polish sausage is made of coarsely ground lean pork with some added beef. The basic spices for this well-known sausage are garlic and marjoram. The recipe can be halved or quartered for smaller portions.

Ingredients

- 8 lbs. pork shoulder or lean trim (75% lean)
- 2 lbs. beef trimmings (80% lean)
- 4.8 oz. ice or water
- 7¼ tablespoons salt
- 4 tablespoons sugar
- 2½ tablespoons white pepper
- 2¼ teaspoons mustard seed
- 4 teaspoons marjoram
- 1½ teaspoons garlic powder
- 2½ teaspoons nutmeg
- 1⅓ cups nonfat dry milk

1. Grind beef and pork through ¼-inch plate. Add spices and water, mix thoroughly. Grind through 3/16-inch plate.
2. Stuff into natural hog casings. Refrigerate or freeze immediately.

Venison or Game Sausage

Venison is a high quality, delicious, and nutritious meat. Care should be used in handling venison just as you would any other meat. Most of the flavor in a meat product is in the fat; therefore, in making a breakfast type sausage using game meat, pork fat is used. An average deer will yield 50 to 60 pounds of venison. This recipe is perfect if

you want to save half your venison for steaks or other use and use half for sausage. If you have a smaller quantity of venison, just divide the recipe accordingly.

Ingredients

- 25 lbs. lean venison or trimmings
- 25 lbs. fat pork (jowls or fresh bellies)
- 2 cups salt
- ⅔ cup black pepper
- ⅓ cup ground ginger
- ½ cup rubbed sage
- ¼ cup crushed red pepper (optional)
- ¼ cup ground red pepper (optional)

1. Cut lean venison and pork into small pieces, add spices, and mix. Grind twice through 1/8-inch or 3/16-inch plate.
2. Sausage may be stuffed or pattied. Refrigerate or freeze immediately.

Cooked Bratwurst

Bratwurst is a typical fresh German style sausage. It is a mild sausage, often served with sauerkraut.

Ingredients

- 10 lbs. pork trim (70% lean)
- 3 lbs. ice or water
- 6½ tablespoons salt
- 1½ tablespoons ground white pepper
- 1¾ teaspoons sugar
- ½ tablespoon mace
- 1 tablespoon onion powder

1. Grind pork through ¼-inch plate and mix with salt, water, and spices. Stuff in natural hog casings or 32-mm collagen casings.
2. Steam or water cook at 170°F to an internal temperature of 155°F. Store in the refrigerator or freezer.

Curing Virginia Ham

Meat has been preserved for centuries by drying, salting, and curing. Virginia ham was one of the first agricultural products exported from North America. Today, Virginia ham is still loved around the world for its distinctive savory taste. To cure ham at home, a few simple rules should be followed.

Start with a Good Ham

To get a high quality cured ham, you have to start with a high quality fresh ham. Choose ham from young, healthy, fast-growing hogs with a desirable lean-to-fat ratio. If you are not butchering your own animal, fresh hams can be purchased from grocery stores or a local butcher. Hams for curing should have a long thick cushion (Figure 1 on page 68), a deep and wide butt face, minimal seam, and external fat as seen on the collar (Figure 1) and alongside the butt face (Figure 3A on page 68), and weigh less than 24 pounds. Heavier hams are normally fatter and are more likely to spoil before the cure adjuncts penetrate to prevent deterioration.

Keep the Hams Properly Chilled

Fresh hams should be kept chilled below 40°F before being cured and then kept between 36° and 40°F during curing.

Cure Application

Hams can be cured in just salt, or you can add a dry sugar cure for a richer, sweeter flavor. For each 25 lbs. of fresh meat, use:

- 2 pounds salt
- ½ pound sugar
- ½ ounce coarse kosher salt

Mix these ingredients thoroughly and divide into two equal parts. Apply the first half on day one and the second portion on

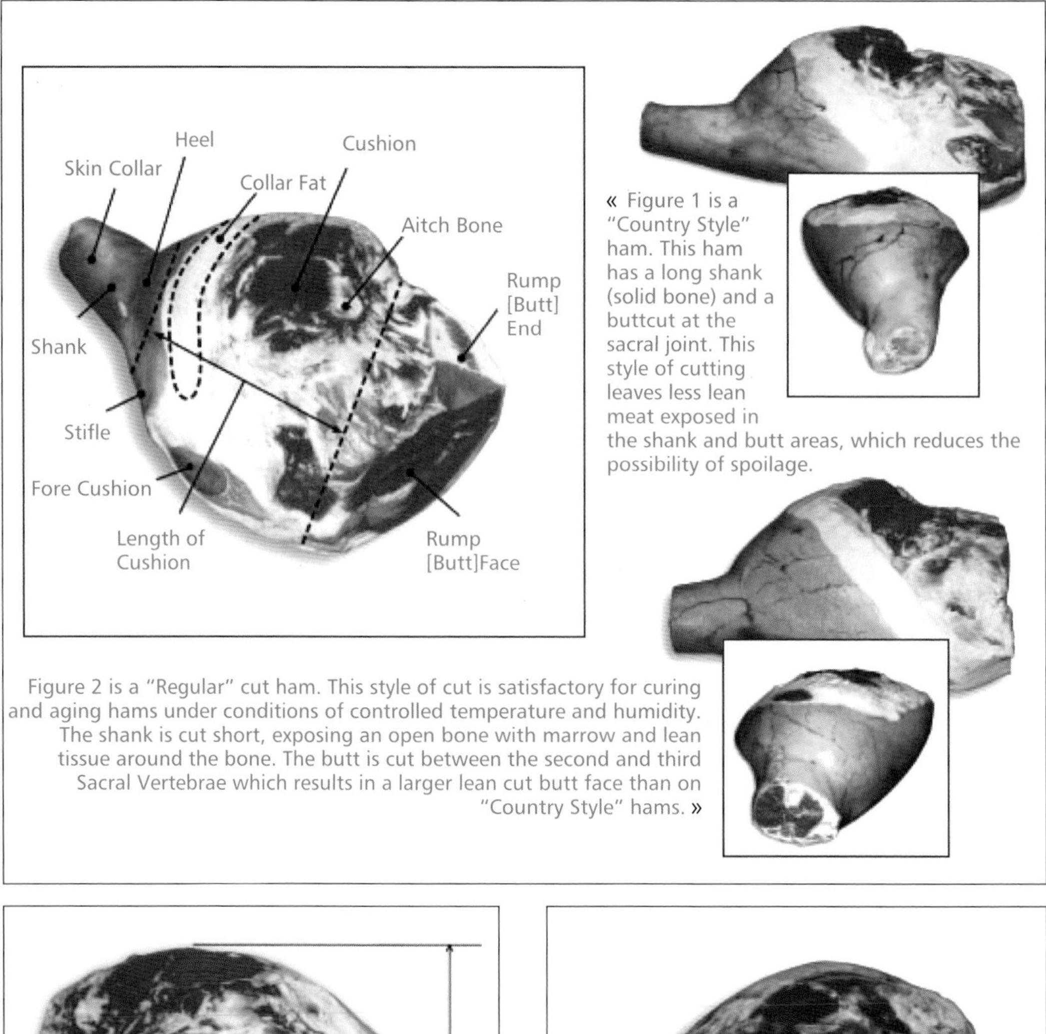

« Figure 1 is a "Country Style" ham. This ham has a long shank (solid bone) and a buttcut at the sacral joint. This style of cutting leaves less lean meat exposed in the shank and butt areas, which reduces the possibility of spoilage.

Figure 2 is a "Regular" cut ham. This style of cut is satisfactory for curing and aging hams under conditions of controlled temperature and humidity. The shank is cut short, exposing an open bone with marrow and lean tissue around the bone. The butt is cut between the second and third Sacral Vertebrae which results in a larger lean cut butt face than on "Country Style" hams. »

Depth of Cushion

Fat alongside the Butt Face

« Figure 3A illustrates a high-quality ham that has a firm, bright-colored lean with at least a small amount of marbling (specks of fat in the lean) in the butt face.

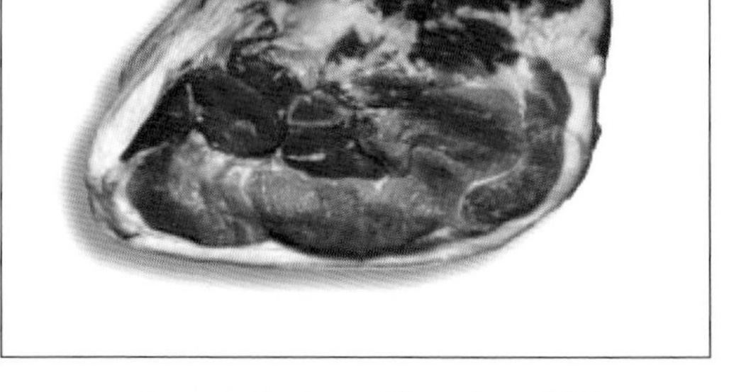

« Figure 3B reveals the type of ham to avoid. Its muscles are soft, usually pale in color and lack marbling. They also "weep" excessively and will shrink more during curing. The open seams between the muscles allow bacterial and insect invasion.

day seven of the curing period. Rub the curing mixture into all lean surfaces (Figure 1) of the ham. Cover the skin and fat (little will be absorbed through these surfaces).

Virginia-style hams should be cured seven days per inch of cushion depth (Figure 3A), or one and a half days per pound of ham. Keep accurate records of placing hams in cure and mark on your calendar when it's time to remove hams from cure. During the curing period, keep hams at a temperature of 36 to 40°F.

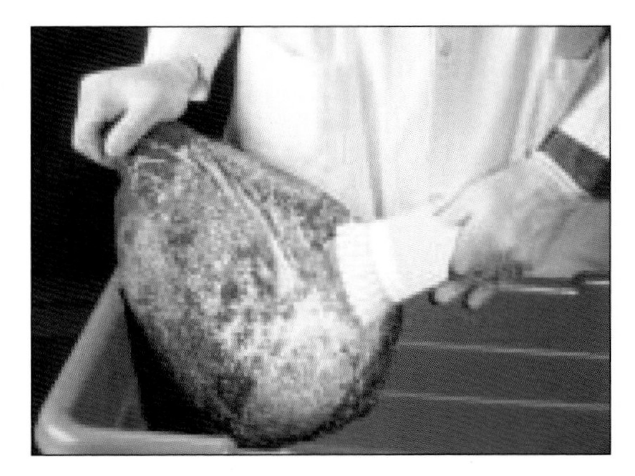

After Curing—Soak and Wash

When the curing period has passed, the hams should be placed in a tub of clean, cold water for one hour. This will dissolve most of the surface curing mix and make the meat receptive to smoke. After soaking, scrub the ham with a stiff bristle brush and allow it to dry.

Cure Equalization

After the ham has been washed, store it at 50-60°F for approximately fourteen days to permit the cure adjuncts to be distributed evenly throughout the ham. The product will shrink approximately 8–10 percent during cure application and equalization.

Final Steps

In Southeastern Virginia, most hams are smoked to accelerate drying and to give added flavor. The Smithfield ham is smoked for a long time at a low temperature (under 90°F). Wood from hardwood species of trees (trees that shed their leaves in the fall) should be used to produce the smoke. Hickory is the most popular, but apple, plum, peach, oak, maple, beech, ash, or cherry may be used. Pine, cedar, spruce, and other "needle leaf" trees are *not* to be used for smoking meat since they give off resin which has a bitter taste and odor.

The fire should be a "cool" smoldering type that produces dense smoke. The temperature of the smokehouse should be kept below 90°F. Hams should be hung in a smokehouse so that they do not touch each other. Smoke until they become chestnut brown in color, which may take one to three days.

In Southwest Virginia, hams are not traditionally smoked after curing, washing, and equalizing. Instead, the final step is to rub hams with the following mixture.

For 25 lbs. of ham:

- ½ pound black pepper
- 1 cup molasses
- ¼ pound of brown sugar
- ¼ ounce coarse kosher salt
- ¼ ounce of cayenne pepper

Then bag the hams as shown on the left.

Age the Hams for 45 to 180 Days

As with wine or cheese, it is during the aging process that rich flavor develops. Age hams for 45 to 180 days at 75 to 95°F and a relative humidity of 55 to 65 percent. Use an exhaust fan controlled by a humidistat to limit mold growth and prevent excessive drying. Air circulation is needed, particularly during the first seven to ten days of aging, to dry the ham surface. Approximately 8 to 12 percent of the initial weight is lost.

Pests

As unpleasant as it is to think about, it's important to know that pests are drawn to cured meat and can ruin all your hard work if you don't take certain precautions. The insects attracted to cured meat are the cheese skipper, larder beetle, and red-legged ham beetle. Mites, which are not officially insects, also may infest cured meats.

1. **Cheese Skipper**

This insect gets its name from the jumping habit of the larvae which bore through cheese and cured meats. Meat infested with this insect quickly rots and becomes slimy. Adult flies are two-winged and are one-third the size of houseflies. They lay their eggs on meat and cheese and multiply rapidly.

2. **Larder Beetle**

This insect is dark brown and has a yellowish band across its back. Its larvae feed on or immediately beneath the cured meat surface, but do not rot the meat. The larvae are fuzzy, brownish, and about 1/3 inch long at maturity.

3. **Red-Legged Ham Beetle**

The larvae are purplish and about ⅓ inch long. They bore through the meat and cause it to dry rot. Adults are about ¼ inch long, brilliant greenish blue with red legs and are red at the bases of their antennae. They feed on the meat surface.

4. **Mites**

Mites are whitish and about 1/32 inch long at maturity. Affected parts of meat infested with mites appear powdery.

To help prevent insect and mite infestation, start the curing and aging during cold weather when these insects are inactive. Proper cleaning of the aging and storage areas is essential since the cheese skipper feeds and breeds on grease and tiny scraps of meat lodged in cracks. Cracks should be sealed with putty or plastic wood after cleaning.

Screens should be installed to prevent entrance of flies, ants, and other insects that carry mites.

Double-entry doors are recommended to reduce infestation of insects.

If any product becomes infested after precautions have been taken, it should be removed from the storeroom and the infested area should be trimmed. The trim should be deep enough to remove larvae that have penetrated along the bone and through the fat. The uninfested portion is safe to eat, but should be prepared and consumed promptly. The exposed lean of the trimmed areas should be protected by greasing it with salad oil or melted fat to delay molding or drying.

Protect the hams by placing a barrier between the meat and the insects. Heavy brown grocery bags with no rips or tears in them are ideal to use for this purpose. Place the ham in a bag and fold and tie the top. Then, place the bagged ham in a second bag, fold and tie. The hams wrapped by this method can be hung in a dry, cool, protected room to age. This room should be clean, tight, and well-ventilated.

Preparing the Ham

The traditional four-step method of preparing Virginia ham is to:

1. Wash ham with a stiff-bristled brush, removing as much of the salt as possible.
2. Place the ham in a large container, cover with cold water, and allow it to stand 10–12 hours or overnight.
3. Lift the ham from the water and place it in a deep kettle with the skin side up and cover with fresh, cold water.
4. Cover the kettle, heat to a boil, but reduce heat as soon as the water boils. Simmer 20 to 25 minutes per pound until done.

Another method of cooking is to soak, scrub, and place the ham in a covered roaster, fat side up. Then, pour 2 inches of water into the roaster and place it in a 325°F oven. Cook approximately 20 to 25 minutes per pound. Baste frequently. Cook to an internal temperature of 155°F as indicated by a meat thermometer placed in the thickest position of the ham cushion. If you do not have a thermometer, test for doneness by moving the flat (pelvic) bone. It should move easily when ham is done. Lift ham from kettle. Remove skin. Sprinkle with brown sugar and/or bread crumbs and brown lightly in a 375°F oven, or use one of the suggested glazes.

Orange Glaze: Mix 1 cup brown sugar, juice, and grated rind of one orange, spread over fat surface. Bake until lightly browned in a 375°F oven. Garnish with orange slices.

Mustard Glaze: Mix ¼ cup brown sugar, 2 teaspoons prepared mustard, 2 tablespoons vinegar, and 1 tablespoon water. Spread over fat surface and bake as directed above.

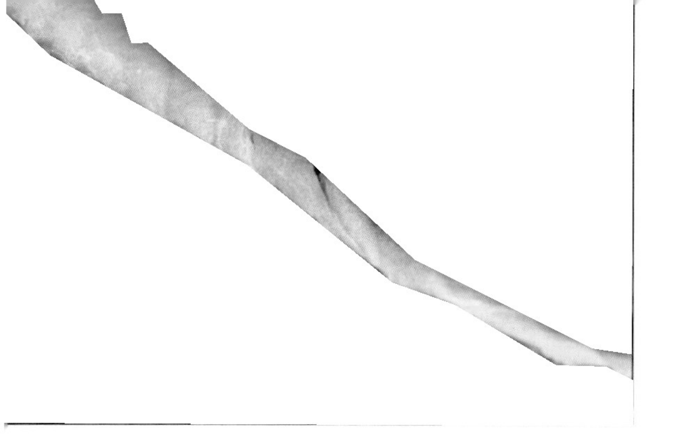

Spice Glaze: Use 1 cup brown sugar and 1 cup juice from spiced peaches or crab apples. Bake as directed above. Garnish with the whole pickled fruit.

Cooking Ham Slices

Baking: Place thick slice in covered casserole and bake in 325°F oven. Brown sugar and cloves, fruit juice, or mustard-seasoned milk may be used over the ham during baking. Uncover the last 15 to 20 minutes for browning.

Broiling: Score fat edges and lay on broiler rack. Place 4 inches from broiler and broil for specified time, turning only once.

Frying: Trim the skin off the ham slices. Cut the outer edge of fat in several places to prevent it from curling during cooking. Place a small amount of fat in a moderately hot skillet. When it has melted, add ham slices. Cook ham slowly, turning often. Allow about 10 minutes total cooking time for thin slices. Remove ham from pan and add a small amount of water to raise the drippings for red-eye gravy. To decrease the salty taste, fry ham with a small amount of water in the skillet.

How To Carve a Ham

The most delightful flavor of Virginia ham can be enjoyed from thin slices. Thus, a very sharp knife, preferably long and narrow, is needed. With the ham on a platter, dressed side up, make a cut perpendicular to the bone about 6 inches in from the end of the hock.

How To Debone Cooked Ham

The ham is easier to slice when the bones are removed while the ham is warm.

- Place skinned ham fat side down on three or four strips of firm white cloth 3 inches wide and long enough to reach around ham and tie. Do not tie until bone is removed.
- Remove flat aitch bone (pelvic) by scalping around it.
- Take sharp knife and, beginning at hock end, cut to bone the length of ham. Follow bones with point of knife as you cut.
- Loosen meat from bones. Remove bones.
- Tie cloth strips together, pulling ham together as you tie.
- Chill in the refrigerator overnight. Slice very thin, or have the ham sliced by machine.

The Home Dairy

Make Your Own Butter

Making butter the old-fashioned way is incredibly simple and very gratifying. It's a great project to do with kids, too. All you need are a jar, a marble, some fresh cream, and about 20 minutes.

1. Start with about twice as much heavy whipping cream as you'll want butter. Pour it into the jar, drop in the marble, close the lid tightly, and start shaking.
2. Check the consistency of the cream every three to four minutes. The liquid will turn into whipped cream, and then eventually you'll see little clumps of butter forming in the jar. Keep shaking for another few minutes and then begin to strain out the liquid into another jar. This is buttermilk, which is great for use in making pancakes, waffles, biscuits, and muffins.
3. The butter is now ready, but it will store better if you wash and work it. Add ½ cup of ice cold water and continue to shake for two or three minutes. Strain out the water and repeat. When the strained water is clear, mash the butter to extract the last of the water, and strain.

4. Scoop the butter into a ramekin, mold, or wax paper.

If desired, add salt or chopped fresh herbs to your butter just before storing or serving. Butter can also be made in a food processor or blender to speed up the processing time.

Make Your Own Yogurt

Yogurt is basically fermented milk. You can make it by adding the active cultures *Streptococcus thermophilus* and *Lactobacillus bulgaricus* to heated milk, which will produce lactic acid, creating yogurt's tart flavor and thick consistency. Yogurt is simple to make and is delicious on its own, as a dessert, in baked goods, or in place of sour cream.

Yogurt is thought to have originated many centuries ago among the nomadic tribes of Eastern Europe and Western Asia. Milk stored in animal skins would acidify and coagulate. The acid helped preserve the milk from further spoilage and from the growth of pathogens (disease-causing microorganisms).

Ingredients

Makes 4 to 5 cups of yogurt

- **1 quart milk** (cream, whole, low-fat, or skim)—In general, the higher the milk fat level in the yogurt, the creamier and smoother it will taste. **Note:** If you use home-produced milk it *must* be pasteurized before preparing yogurt. See box at the top of page 74 for tips on pasteurizing milk.
- **Nonfat dry milk powder**—Use ⅓ cup powder when using whole or low-fat milk, or use ⅔ cup powder when using skim milk. The higher the milk solids, the firmer the yogurt will be. For even more firmness, add gelatin (directions below).
- **Commercial, unflavored, cultured yogurt**—Use ¼ cup. Be sure the product label indicates that it contains a live culture. Also note the content of the culture. *L. bulgaricus* and *S. thermophilus* are required in yogurt, but some manufacturers may in addition

add *L. acidophilus* or *B. bifidum*. The latter two are used for slight variations in flavor, but more commonly for health benefits attributed to these organisms. All culture variations will make a successful yogurt.

- **2 to 4 tablespoons sugar or honey (optional)**
- **1 teaspoon unflavored gelatin (optional)**—For a thick, firm yogurt, swell 1 teaspoon gelatin in a little milk for 5 minutes. Add this to the milk and nonfat dry milk mixture before cooking.

HOW TO PASTEURIZE RAW MILK

If you are using fresh milk that hasn't been processed, you can pasteurize it yourself. Heat water in the bottom section of a double boiler and pour milk into the top section. Cover the milk and heat to 165°F while stirring constantly for uniform heating. Cool immediately by setting the top section of the double boiler in ice water or cold running water. Store milk in the refrigerator in clean containers until ready for making yogurt.

Supplies

- **Double boiler or regular saucepan**—1 to 2 quarts in capacity larger than the volume of yogurt you wish to make.
- **Cooking or jelly thermometer**—A thermometer that can clip to the side of the saucepan and remain in the milk works best. Accurate temperatures are critical for successful processing.
- **Mixing spoon**
- **Yogurt containers**—cups with lids or canning jars with lids.
- **Incubator**—a yogurt-maker, oven, heating pad, or warm spot in your kitchen. To use your oven, place yogurt containers into deep pans of 110°F water. Water should come at least halfway up the containers. Set oven temperature at lowest point to maintain water temperature at 110°F. Monitor temperature throughout incubation, making adjustments as necessary.

Processing

1. **Combine ingredients and heat.** Heating the milk is necessary in order to change the milk proteins so that they set together rather than form curds and whey. Do not substitute this heating step for pasteurization. Place cold, pasteurized milk in top of a double boiler and stir in nonfat dry milk powder. Adding nonfat dry milk to heated milk will cause some milk proteins to coagulate and form strings. Add sugar or honey if a sweeter, less tart yogurt is desired. Heat milk to 200°F, stirring gently, and hold for 10 minutes for thinner yogurt, or hold 20 minutes for thicker yogurt. Do not boil. Be careful and stir constantly to avoid scorching if not using a double boiler.
2. **Cool and inoculate.** Place the top of the double boiler in cold water to cool milk rapidly to 112 to 115°F. Remove 1 cup of the warm milk and blend it with the yogurt starter culture. Add this to the rest of the warm milk. The temperature of the mixture should now be 110 to 112°F.
3. **Incubate.** Pour immediately into clean, warm containers; cover and place in prepared incubator. Close the incubator and incubate about 4 to 7 hours at 110°F, ± 5°F. Yogurt should set firm when the proper acid level is achieved (pH 4.6). Incubating yogurt for several hours past the time after the yogurt has set will produce more acidity. This will result in a more tart or acidic flavor and eventually cause the whey to separate.
4. **Refrigerate.** Rapid cooling stops the development of acid. Yogurt will keep for about ten to twenty-one days if held in the refrigerator at 40°F or lower.

Yogurt Types

Set yogurt: A solid set where the yogurt firms in a container and is not disturbed.

Stirred yogurt: Yogurt made in a large container then spooned or otherwise dispensed into secondary serving containers. The consistency of the "set" is broken and the texture is less firm than set yogurt. This is the most popular form of commercial yogurt.

Drinking yogurt: Stirred yogurt into which additional milk and flavors are mixed. Add fruit or fruit syrups to taste. Mix in milk to achieve the desired thickness. The shelf life of this product is four to ten days, since the pH is raised by the addition of fresh milk. Some whey separation will occur and is natural. Commercial products recommend a thorough shaking before consumption.

Fruit yogurt: Fruit, fruit syrups, or pie filling can be added to the yogurt. Place them on top, on bottom, or stir them into the yogurt.

Troubleshooting

- If milk forms some clumps or strings during the heating step, some milk proteins may have jelled. Take the solids out with a slotted spoon or, in difficult cases, after cooking, pour the milk mixture through a clean colander or cheesecloth before inoculation.
- When yogurt fails to coagulate (set) properly, it's because the pH is not low enough. Milk proteins will coagulate when the pH has dropped to 4.6. This is done by the culture growing and producing acids. Adding culture to very hot milk (+115°F) can kill bacteria. Use a thermometer to carefully control temperature.
- If yogurt takes too long to make, it may be because the temperature is off. Too hot or too cold of an incubation temperature can slow down culture growth. Use a thermometer to carefully control temperature.
- If yogurt just isn't working, it may be because the starter culture was of poor quality. Use a fresh, recently purchased culture from the grocery store each time you make yogurt.
- If yogurt tastes or smells bad, it's likely because the starter culture is contaminated. Obtain new culture for the next batch.
- Yogurt has overset or incubated too long. Refrigerate yogurt immediately after a firm coagulum has formed.
- If yogurt tastes a little odd, it could be due to overheating or boiling of the milk. Use a thermometer to carefully control temperature.
- When whey collects on the surface of the yogurt, it's called syneresis. Some syneresis is natural. Excessive separation of whey, however, can be caused by incubating yogurt too long or by agitating the yogurt while it is setting.

Storing Your Yogurt

- Always pasteurize milk or use commercially pasteurized milk to make yogurt.
- Discard batches that fail to set properly, especially those due to culture errors.
- Yogurt generally has a ten- to twenty-one-day shelf life when made and stored properly in the refrigerator below 40°F.
- Always use clean and sanitized equipment and containers to ensure a long shelf life for your yogurt. Clean equipment and containers in hot water with detergent, then rinse well. Allow to air dry.

Make Your Own Cheese

There are endless varieties of cheese you can make, but they all fall into two main categories: soft and hard. Soft cheeses (like cream cheese) are easier to make because they don't require a cheese press. The curds in hard cheeses (like cheddar) are pressed

together to form a solid block or wheel, which requires more time and effort, but hard cheeses will keep longer than soft cheeses, and generally have a much stronger flavor.

Cheese is basically curdled milk and is made by adding an enzyme (typically rennet) to milk, allowing curds to form, heating the mixture, straining out the whey, and finally pressing the curds together. Cheeses such as *queso fresco* or *queso blanco* (traditionally eaten in Latin American countries) and *paneer* (traditionally eaten in India) are made with an acid such as vinegar or lemon juice instead of bacterial cultures or rennet.

You can use any kind of milk to make cheese, including cow's milk, goat's milk, sheep's milk, and even buffalo's milk (used for traditional mozzarella). For the richest flavor, try to get raw milk from a local farmer. If you don't know of one near you, visit www.realmilk.com/where.html for a listing of raw milk suppliers in your state. You can use homogenized milk, but it will produce weaker curds and a milder flavor. If your milk is pasteurized, you'll need to "ripen" it by heating it in a double boiler until it reaches 86°F and then adding 1 cup of unpasteurized, preservative-free cultured buttermilk per gallon of milk and letting it stand 30 minutes to three hours (the longer you leave it, the sharper the flavor will be). If you cannot find unpasteurized buttermilk, diluting ⅛ teaspoon calcium chloride (available from online cheesemaker suppliers) in ¼ cup of water and adding it to your milk will create a similar effect.

Rennet (also called rennin or chymosin) is sold online at cheesemaking sites in tablet or liquid form. You may also be able to find Junket rennet tablets near the pudding and gelatin in your grocery store. One teaspoon of liquid rennet is the equivalent of one rennet tablet, which is enough to turn 5 gallons of milk into cheese (estimate four drops of liquid rennet per gallon of milk). Microbial rennet is a vegetarian alternative that is available for purchase online.

Preparation

It's important to keep your hands clean and all equipment sterile when making cheese.

1. Wash hands and all equipment with soapy detergent before and after use.
2. Rinse all equipment with clean water, removing all soapy residue.
3. Boil all cheesemaking equipment between uses.
4. For best quality cheese, use new cheesecloth each time you make cheese. (Sterilize cheesecloth by first washing, then boiling.)
5. Squeaky clean is clean. If you can feel a residue on the equipment, it is not clean.

Yogurt Cheese

This soft cheese has a flavor similar to sour cream and a texture like cream cheese. A pint of yogurt will yield approximately ¼ pound of cheese. The yogurt cheese has a

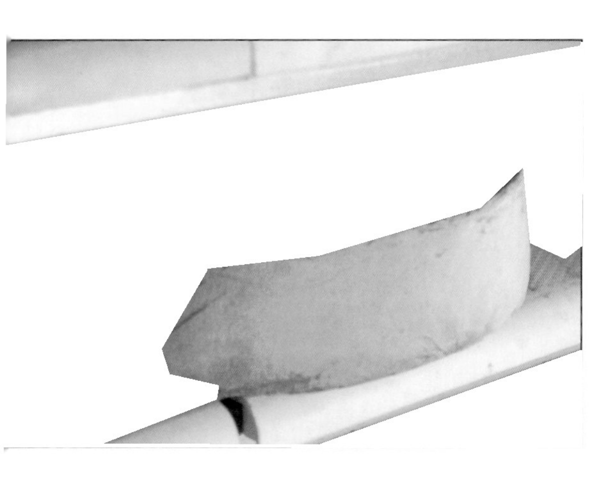

shelf life of approximately seven to fourteen days when wrapped and placed in the refrigerator and kept at less than 40°F. Add a little salt and pepper and chopped fresh herbs for variety.

1. Line a large strainer or colander with cheesecloth.
2. Place the lined strainer over a bowl and pour in plain, whole-milk yogurt. Do not use yogurt made with the addition of gelatin, as gelatin will inhibit whey separation.
3. Let yogurt drain overnight, covered with plastic wrap. Empty the whey from the bowl.
4. Fill a strong plastic storage bag with some water, seal, and place over the cheese to weigh it down. Let the cheese stand another eight hours and then enjoy!

Queso Blanco

Queso blanco is a white, semihard cheese made without culture or rennet. It is eaten fresh and may be flavored with peppers, herbs, and spices. It is considered a "frying cheese," meaning it does not melt and may be deep-fried or grilled. *Queso blanco* is best eaten fresh, so try this small recipe the first time you make it. If it disappears quickly, next time double or triple the recipe. This recipe will yield about ½ cup of cheese.

- 2 cups milk
- 4 teaspoons white vinegar
- Salt
- Minced jalapeño, black pepper, chives, or other herbs to taste

1. Heat milk to 176°F for 20 minutes.
2. Add vinegar slowly to the hot milk until the whey is semiclear and the curd particles begin to form stretchy clumps. Stir for 5 to 10 minutes. When it's ready, you should be able to stretch a piece of curd about 1/3 inch before it breaks.
3. Allow to cool, and strain off the whey by filtering through a cheesecloth-lined colander or a cloth bag.
4. Work in salt and spices to taste.
5. Press the curd in a mold or simply leave in a ball.
6. *Queso blanco* may keep for several weeks if stored in a refrigerator, but is best eaten fresh.

Ricotta Cheese

Making ricotta is very similar to making *queso blanco*, though it takes a bit longer. Start the cheese in the morning for use at dinner, or make a day ahead. Use it in lasagna, in desserts, or all on its own.

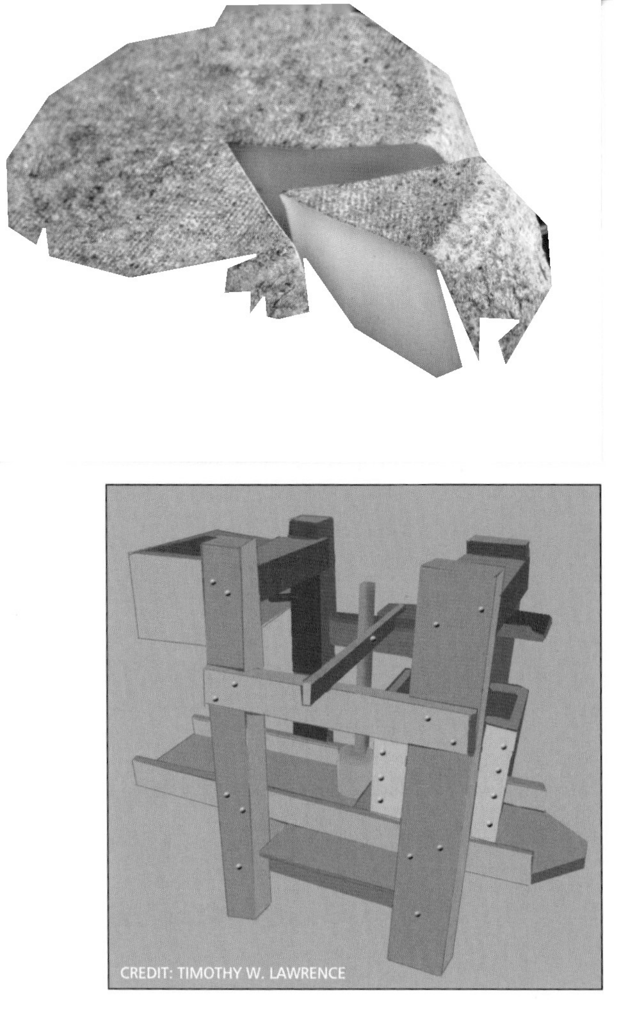

⮝ This press works for both making cider and for pressing cheese. The apple press part is a scissors jack (found at an auto parts store) mounted to one of the top timbers. The opposite end is a grinder, made up of two oak rollers. Stainless steel screws serve as teeth to mash the apples, which are then strained in the mesh-lined bucket. You can then press the apples on the opposite end.

The cheese press is a wooden arm mounted across the two top timbers. Another arm goes straight down into the cheese mold. A water-filled jug is hung from the end. The pressure is varied by adding or subtracting water from the jug.

- 1 gallon milk
- ¼ teaspoon salt
- ⅓ cup plus 1 teaspoon white vinegar

1. Pour milk into a large pot, add salt, and heat slowly while stirring until the milk reaches 180°F.
2. Remove from heat and add vinegar. Stir for 1 minute as curds begin to form.
3. Cover and allow to sit undisturbed for 2 hours.

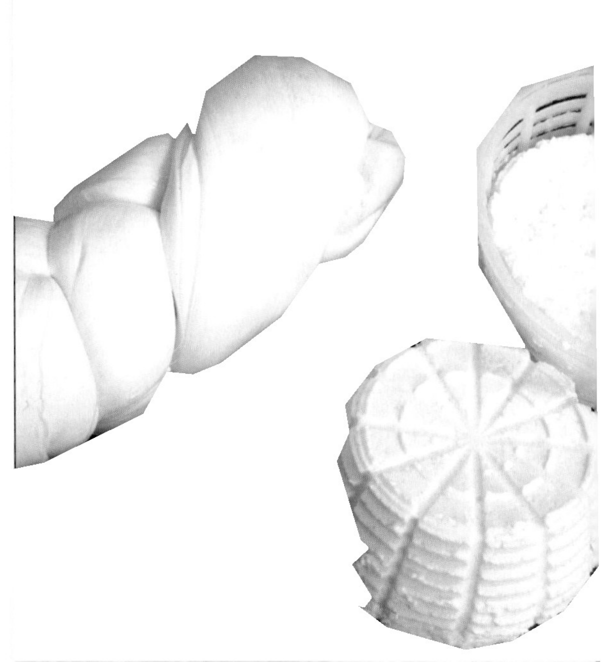

4. Pour mixture into a colander lined with cheesecloth, and allow to drain for 2 or more hours.
5. Store in a sealed container for up to a week.

Mozzarella

This mild cheese will make your homemade pizza especially delicious. Or slice it and eat with fresh tomatoes and basil from the garden. Fresh cheese can be stored in salt water but must be eaten within two days.

- 1 gallon 2% milk
- ¼ cup fresh, plain yogurt (see recipe on page 150)
- One tablet rennet or 1 teaspoon liquid rennet dissolved in ½ cup tap water
- Brine: use 2 pounds of salt per gallon of water

1. Heat milk to 90°F and add yogurt. Stir slowly for 15 minutes while keeping the temperature constant.
2. Add rennet mixture and stir for 3 to 5 minutes.

MAKE YOUR OWN SIMPLE CHEESE PRESS

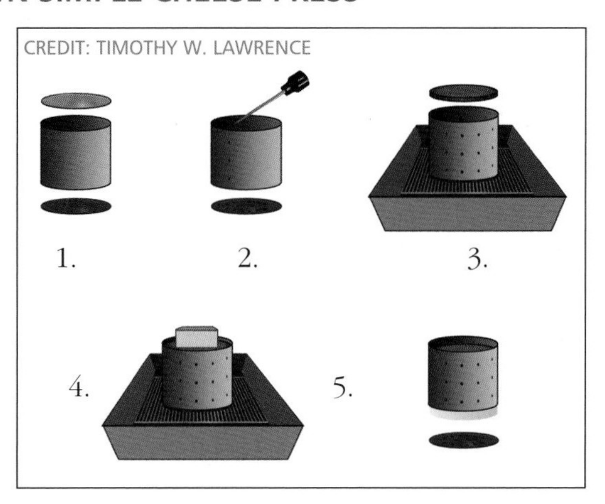

1. Remove both ends of a large coffee can or thoroughly cleaned paint can, saving one end.
2. Use an awl or a hammer and long nail to pierce the sides in several places, piercing from the inside out.
3. Place the can on a cooling rack inside a larger basin. Leave the bottom of the can in place.
4. Use a saw to cut a ¾-inch-thick circle of wood to create a "cheese follower." It should be small enough in diameter to fit easily in the can. Place cheese curds in the can, and top with the cheese follower. Place several bricks wrapped in cloth or foil on top of the cheese follower to weigh down curds.
5. Once the cheese is fully pressed, remove the bricks and bottom of the can. Use the cheese follower to push the cheese out of the can.

3. Cover, remove from heat, and allow to stand until coagulated, about 30 minutes.
4. Cut curd into ½-inch cubes. Allow to stand for 15 minutes with occasional stirring.
5. Return to heat and slowly increase temperature to 118°F over a period of 45 minutes. Hold this temperature for an additional 15 minutes.
6. Drain off the whey by transferring the mixture to a cheesecloth-lined colander. Use a spoon to press the liquid out of the curds. Transfer the mat of curd to a flat pan that can be kept warm in a low oven. Do not cut mat, but turn it over every 15 minutes for a 2-hour period. Mat should be tight when finished.
7. Cut the mat into long strips 1 to 2 inches wide and place in hot water (180°F). Using wooden spoons, tumble and stretch it under water until it becomes elastic, about 15 minutes.
8. Remove curd from hot water and shape it by hand into a ball or a loaf, kneading in the salt. Place cheese in cold water (40°F) for approximately 1 hour.
9. Store in a solution of 2 teaspoons salt to 1 cup water.

Cheddar Cheese

Cheddar is a New England and Wisconsin favorite. The longer you age it, the sharper the flavor will be. Try a slice with a wedge of homemade apple pie.

Ingredients

- 1 gallon milk
- ¼ cup buttermilk
- 1 tablet rennet, or 1 teaspoon liquid rennet
- 1½ teaspoons salt

Directions

1. Combine milk and buttermilk and allow the mixture to ripen overnight.
2. The next day, heat milk to 90°F in a double boiler and add rennet.
3. After about 45 minutes, cut curds into small cubes and let sit 15 minutes.
4. Heat very slowly to 100°F and cook for about an hour or until a cooled piece

of curd will keep its shape when squeezed.

5. Drain curds and rinse out the double boiler.
6. Place a rack lined with cheesecloth inside the double boiler and spread the curds on the cloth. Cover and reheat at about 98°F for 30 to 40 minutes. The curds will become one solid mass.
7. Remove the curds, cut them into 1-inch wide strips, and return them to the pan. Turn the strips every 15 to 20 minutes for one hour.
8. Cut the strips into cubes and mix in salt.
9. Let the curds stand for 10 minutes, place them in cheesecloth, and press in a cheese press with 15 pounds for 10 minutes, then with 30 pounds for an hour.
10. Remove the cheese from the press, unwrap it, dip in warm water, and fill in any cracks.
11. Wrap again in cheesecloth and press with 40 pounds for twenty-four hours.
12. Remove from the press and let the cheese dry about five days in a cool, well-ventilated area, turning the cheese twice a day and wiping it with a clean cloth. When a hard skin has formed, rub with oil or seal with wax. You can eat the cheese after six weeks, but for the strongest flavor, allow cheese to age for six months or more.

Make Your Own Ice Cream

Supplies

- 1-pound coffee can
- 3-pound coffee can
- Duct tape
- Ice
- 1 cup salt

HOMEMADE CRACKERS

Crackers are very easy to make and can be varied endlessly by adding seasonings of your choice. Try sprinkling a coarse sea salt and dried oregano or cinnamon and sugar over the crackers just before baking. Serve with homemade cheese.

- 1½ cups all-purpose flour
- 1½ cups whole wheat flour
- 1 teaspoon salt
- 1 cup warm water
- ⅓ cup olive oil
- Herbs, spices, or coarse sea salt as desired

1. Stir together the dry ingredients in a mixing bowl. Add the water and olive oil and knead until dough is elastic and not too sticky (about 5 minutes in an electric mixer with a dough attachment or 10 minutes by hand).
2. Allow dough to rest at room temperature for about half an hour. Preheat oven to 450°F.
3. Flour a clean, dry surface and roll dough to about 1/8 inch thick. Cut into squares and place on a cookie sheet. Sprinkle with desired topping. Bake for about 5 minutes or until crackers are golden brown.

Ingredients

- 2 cups half and half
- ½ cup sugar
- 1 teaspoon vanilla

Directions

1. Mix all the ingredients in the 1-pound coffee can. Cover the lid with duct tape to ensure it is tightly sealed.
2. Place the smaller can inside the larger can and fill the space between the two with ice and salt.
3. Cover the large can and seal with duct tape. Roll the can back and forth for 15 minutes. To reduce noise, place a towel on your working surface, or work on a rug.
4. Dump out ice and water. Stir contents of small can. Store ice cream in a glass or plastic container (if you leave it in the can it may take on a metallic flavor).

If desired, add cocoa powder, coffee granules, crushed peppermint sticks or other candy, or fruit.

Canning

Canning began in France, at the turn of the nineteenth century, when Napoleon Bonaparte was desperate for a way to keep his troops well fed while on the march. In 1800, he decided to hold a contest, offering 12,000 francs to anyone who could devise a suitable method of food preservation. Nicolas François Appert, a French confectioner, rose to the challenge, considering that if wine could be preserved in bottles, perhaps food could be as well. He experimented until he was able to prove that heating food to boiling after it had been sealed in airtight glass bottles prevented the food from deteriorating. Interestingly, this all took place about 100 years before Louis Pasteur found that heat could destroy bacteria. Nearly ten years after the contest began, Napoleon personally presented Appert with the cash reward.

Canning practices have evolved over the last two centuries, but the principles remain the same. In fact, the way we can foods today is basically the same way our grandparents and great-grandparents preserved their harvests for the winter months.

On the next few pages, you will find descriptions of proper canning methods, with details on how canning works and why it is both safe and economical. Much of the information here is from the USDA, which has done extensive research on home canning and preserving. If you are new to home canning, read this section carefully as it will help to ensure success with the recipes that follow.

Whether you are a seasoned home canner or this is your first foray into food preservation, it is important to follow directions carefully. With some recipes, it is okay to experiment with varied proportions or added ingredients, and with others it is important to stick to what's written. In many instances, it is noted whether or not creative liberty is a good idea for a particular recipe, but if you are not sure, play it safe—otherwise you may end up with a jam that is too runny, a vegetable that is mushy, or a product that is spoiled. Take time to read the directions and prepare your foods and equipment adequately and you will find that home canning is safe, economical, tremendously satisfying, and a great deal of fun!

The Benefits of Canning

Canning is fun, economical, and a good way to preserve your precious produce. As more and more farmers' markets make their way into urban centers, city dwellers are also discovering how rewarding it is to make seasonal treats last all year round. Besides the value of your labor, canning home-grown or locally grown food may save you half the cost of buying commercially canned food. Freezing food may be simpler, but most people have limited freezer space, whereas cans of food can be stored almost anywhere. And, what makes a nicer, more thoughtful gift than a jar of homemade jam, tailored to match the recipient's favorite fruits and flavors?

The nutritional value of home canning is an added benefit. Many vegetables begin to lose their vitamins as soon as they are harvested. Nearly half the vitamins may be lost within a few days unless the fresh produce is kept cool or preserved. Within one to two weeks, even refrigerated produce loses half or more of certain vitamins. The heating process during canning destroys from one-third to one-half of vitamins A and C, thiamin, and riboflavin. Once canned, foods may lose from 5 to 20 percent of these sensitive vitamins each year. The amounts of other vitamins, however, are only slightly lower in canned compared with fresh food. If vegetables are handled properly and canned promptly after harvest, they can be more nutritious than fresh produce sold in local stores.

The advantages of home canning are lost when you start with poor quality foods, when jars fail to seal properly, when food spoils, and when flavors, texture, color, and nutrients deteriorate during prolonged storage. The tips that follow explain many of these problems and recommend ways to minimize them.

How Canning Preserves Foods

The high percentage of water in most fresh foods makes them very perishable. They spoil or lose their quality for several reasons:

- Growth of undesirable microorganisms—bacteria, molds, and yeasts
- Activity of food enzymes
- Reactions with oxygen
- Moisture loss

TIP

A large stockpot with a lid can be used in place of a boiling-water canner for high-acid foods like tomatoes, pickles, apples, peaches, and jams. Simply place a rack inside the pot so that the jars do not rest directly on the bottom of the pot.

Microorganisms live and multiply quickly on the surfaces of fresh food and on the inside of bruised, insect-damaged, and diseased food. Oxygen and enzymes are present throughout fresh food tissues.

Proper canning practices include:

- Carefully selecting and washing fresh food
- Peeling some fresh foods
- Hot packing many foods
- Adding acids (lemon juice, citric acid, or vinegar) to some foods
- Using acceptable jars and self-sealing lids
- Processing jars in a boiling-water or pressure canner for the correct amount of time

Collectively, these practices remove oxygen; destroy enzymes; prevent the growth of undesirable bacteria, yeasts, and molds; and help form a high vacuum in jars. High vacuums form tight seals, which keep liquid in and air and microorganisms out.

Canning Glossary

Acid foods—Foods that contain enough acid to result in a pH of 4.6 or lower. Includes most tomatoes; fermented and pickled vegetables; relishes; jams, jellies, and marmalades; and all fruits except figs. Acid foods may be processed in boiling water.

Ascorbic acid—The chemical name for vitamin C; commonly used to prevent browning of peeled, light-colored fruits and vegetables.

Blancher—A 6- to 8-quart lidded pot designed with a fitted, perforated basket to hold food in boiling water or with a fitted rack to steam foods. Useful for loosening skins on fruits to be peeled or for heating foods to be hot packed.

Boiling-water canner—A large, standard-sized, lidded kettle with jar rack designed for heat-processing seven quarts or eight to nine pints in boiling water.

Botulism—An illness caused by eating a toxin produced by growth of *Clostridium botulinum* bacteria in moist, low-acid food containing less than 2 percent oxygen and stored between 40 and 120°F. Proper heat processing destroys this bacterium in canned food. Freezer temperatures inhibit its growth in frozen food. Low moisture controls its growth in dried food. High oxygen controls its growth in fresh foods.

Canning—A method of preserving food that employs heat processing in airtight, vacuum-sealed containers so that food can be safely stored at normal home temperatures.

Canning salt—Also called pickling salt. It is regular table salt without the anti-caking or iodine additives.

Citric acid—A form of acid that can be added to canned foods. It increases the acidity of low-acid foods and may improve their flavor.

Cold pack—Canning procedure in which jars are filled with raw food. "Raw pack" is the preferred term for describing this practice. "Cold pack" is often used incorrectly to refer to foods that are open-kettle canned or jars that are heat-processed in boiling water.

Enzymes—Proteins in food that accelerate many flavor, color, texture, and nutritional changes, especially when food is cut, sliced, crushed, bruised, or exposed to air. Proper

blanching or hot-packing practices destroy enzymes and improve food quality.

Exhausting—Removing air from within and around food and from jars and canners. Exhausting or venting of pressure canners is necessary to prevent botulism in low-acid canned foods.

Headspace—The unfilled space above food or liquid in jars that allows for food expansion as jars are heated and for forming vacuums as jars cool.

Heat processing—Treatment of jars with sufficient heat to enable storing food at normal home temperatures.

Hermetic seal—An absolutely airtight container seal that prevents reentry of air or microorganisms into packaged foods.

Hot pack—Heating of raw food in boiling water or steam and filling it hot into jars.

Low-acid foods—Foods that contain very little acid and have a pH above 4.6. The acidity in these foods is insufficient to prevent the growth of botulism bacteria. Vegetables, some varieties of tomatoes, figs, all meats, fish, seafood, and some dairy products are low-acid foods. To control all risks of botulism, jars of these foods must be either heat processed in a pressure canner or acidified to a pH of 4.6 or lower before being processed in boiling water.

Microorganisms—Independent organisms of microscopic size, including bacteria, yeast, and mold. In a suitable environment, they grow rapidly and may divide or reproduce every ten to thirty minutes. Therefore, they reach high populations very quickly. Microorganisms are sometimes intentionally added to ferment foods, make antibiotics, and for other reasons. Undesirable microorganisms cause disease and food spoilage.

Mold—A fungus-type microorganism whose growth on food is usually visible and colorful. Molds may grow on many foods, including acid foods like jams and jellies and canned fruits. Recommended heat processing and sealing practices prevent their growth on these foods.

Mycotoxins—Toxins produced by the growth of some molds on foods.

Open-kettle canning—A non-recommended canning method. Food is heat-processed in a covered kettle, filled while hot into sterile jars, and then sealed. Foods canned this way have low vacuums or too much air, which permits rapid loss of quality in foods. Also, these foods often spoil because they become recontaminated while the jars are being filled.

Pasteurization—Heating food to temperatures high enough to destroy disease-causing microorganisms.

pH—A measure of acidity or alkalinity. Values range from 0 to 14. A food is neutral when its pH is 7.0. Lower values are increasingly more acidic; higher values are increasingly more alkaline.

Pressure canner—A specifically designed metal kettle with a lockable lid used for heat processing low-acid food. These canners have jar racks, one or more safety devices, systems for exhausting air, and a way to measure or control pressure. Canners with 20- to 21-quart capacity are common. The minimum size of canner that should be used has a 16-quart capacity and can hold seven one-quart jars. Use of pressure saucepans with a capacity of less than 16 quarts is not recommended.

PSIG—Pounds per square inch of pressure as measured by a gauge.

Raw pack—The practice of filling jars with raw, unheated food. Acceptable for canning low-acid foods, but allows more rapid quality losses in acid foods that are heat-processed in boiling water. Also called "cold pack."

Style of pack—Form of canned food, such as whole, sliced, piece, juice, or sauce. The term may also be used to specify whether food is filled raw or hot into jars.

Vacuum—A state of negative pressure that reflects how thoroughly air is removed from within a jar of processed food; the higher the vacuum, the less air left in the jar.

Proper Canning Practices

Growth of the bacterium *Clostridium botulinum* in canned food may cause botulism—a deadly form of food poisoning. These bacteria exist either as spores or as vegetative cells. The spores, which are comparable to plant seeds, can survive harmlessly in soil and water for many years. When ideal conditions exist for growth, the spores produce vegetative cells, which multiply rapidly and may produce a deadly toxin within three to four days in an environment consisting of:

- A moist, low-acid food,
- A temperature between 40 and 120°F, and
- Less than 2 percent oxygen.

Botulinum spores are on most fresh food surfaces. Because they grow only in the absence of air, they are harmless on fresh foods. Most bacteria, yeasts, and molds are difficult to remove from food surfaces. Washing fresh food reduces their numbers only slightly. Peeling root crops, underground stem crops, and tomatoes reduces their numbers greatly. Blanching also helps, but the vital controls are the method of canning and use of the recommended research-based processing times. These processing times ensure destruction of the largest expected number of heat-resistant microorganisms in home-canned foods.

Properly sterilized canned food will be free of spoilage if lids seal and jars are stored below 95°F. Storing jars at 50 to 70°F enhances retention of quality.

Food Acidity and Processing Methods

Whether food should be processed in a pressure canner or boiling-water canner to control botulism bacteria depends on the acidity in the food. Acidity may be natural, as in most fruits, or added, as in pickled food. Low-acid canned foods contain too little acidity to prevent the growth of these bacteria. Other foods may contain enough acidity to block their growth or to destroy them rapidly when heated. The term "pH" is a measure of acidity: the lower its value, the more acidic the food. The acidity level in foods can be increased by adding lemon juice, citric acid, or vinegar.

Low-acid foods have pH values higher than 4.6. They include red meats, seafood, poultry, milk, and all fresh vegetables except for most tomatoes. Most products that are mixtures of low-acid and acid foods also have pH values above 4.6 unless their ingredients include enough lemon juice, citric acid, or vinegar to make them acid foods. Acid foods have a pH of 4.6 or lower. They include fruits, pickles, sauerkraut, jams, jellies, marmalade, and fruit butters.

Although tomatoes usually are considered an acid food, some are now known to have pH values slightly above 4.6. Figs also have pH values slightly above 4.6. Therefore, if they are to be canned as acid foods, these products must be acidified to a pH of 4.6 or lower with lemon juice or citric acid. Properly acidified tomatoes and figs are acid foods and can be safely processed in a boiling-water canner.

Botulinum spores are very hard to destroy at boiling-water temperatures; the higher the canner temperature, the more easily they are destroyed. Therefore, all low-acid foods should be sterilized at temperatures of 240 to 250°F, attainable with pressure canners operated at 10 to 15 PSIG. (PSIG means pounds per square inch of pressure as measured by a gauge.) At these temperatures, the time needed to destroy bacteria in low-acid canned foods ranges from twenty to 100 minutes. The exact time depends on the kind of food being canned, the way it is packed into jars, and the size of jars. The time needed to safely process low-acid foods in boiling water ranges from seven to eleven hours; the time needed to process acid foods in boiling water varies from five to eighty-five minutes.

Know Your Altitude

It is important to know your approximate elevation or altitude above sea level in order to determine a safe processing time for canned foods. Since the boiling temperature of liquid is lower at higher elevations, it is critical that additional time be given for the safe processing of foods at altitudes above sea level.

What Not to Do

Open-kettle canning and the processing of freshly filled jars in conventional ovens, microwave ovens, and dishwashers are not recommended because these practices do not prevent all risks of spoilage. Steam canners are not recommended because processing times for use with current models have not been adequately researched. Because steam canners may not heat foods in the same manner as boiling-water canners, their use with boiling-water processing times may result in spoilage. So-called canning powders are useless as preservatives and do not replace the need for proper heat processing.

It is not recommended that pressures in excess of 15 PSIG be applied when using new pressure-canning equipment.

Ensuring High Quality Canned Foods

Examine food carefully for freshness and wholesomeness. Discard diseased and moldy food. Trim small diseased lesions or spots from food.

Can fruits and vegetables picked from your garden or purchased from nearby producers when the products are at their peak of quality—within six to twelve hours after harvest for most vegetables. However, apricots, nectarines, peaches, pears, and plums should be ripened one or more days between harvest and canning. If you must delay the canning of other fresh produce, keep it in a shady, cool place.

Fresh, home-slaughtered red meats and poultry should be chilled and canned without delay. Do not can meat from sickly or diseased animals. Put fish and seafood on ice after harvest, eviscerate immediately, and can them within two days.

Maintaining Color and Flavor in Canned Food

To maintain good natural color and flavor in stored canned food, you must:

- Remove oxygen from food tissues and jars,
- Quickly destroy the food enzymes, and
- Obtain high jar vacuums and airtight jar seals.

Follow these guidelines to ensure that your canned foods retain optimal colors and flavors during processing and storage:

- Use only high quality foods that are at the proper maturity and are free of diseases and bruises
- Use the hot-pack method, especially with acid foods to be processed in boiling water
- Don't unnecessarily expose prepared foods to air; can them as soon as possible
- While preparing a canner load of jars, keep peeled, halved, quartered, sliced or diced apples, apricots, nectarines, peaches, and pears in a solution of 3 grams (3,000 milligrams) ascorbic acid to 1 gallon of cold water. This procedure is also useful in maintaining the natural color of mushrooms and potatoes and for preventing stem-end discoloration in cherries and grapes.

You can get ascorbic acid in several forms:

Pure powdered form—Seasonally available among canning supplies in supermarkets. One level teaspoon of pure powder weighs about 3 grams. Use 1 teaspoon per gallon of water as a treatment solution.

Vitamin C tablets—Economical and available year-round in many stores. Buy 500-milligram tablets; crush and dissolve six tablets per gallon of water as a treatment solution.

Commercially prepared mixes of ascorbic and citric acid—Seasonally available among canning supplies in supermarkets. Sometimes citric acid powder is sold in supermarkets, but it is less effective in controlling discoloration. If you choose to use these products, follow the manufacturer's directions.

- Fill hot foods into jars and adjust headspace as specified in recipes
- Tighten screw bands securely, but if you are especially strong, not as tightly as possible
- Process and cool jars
- Store the jars in a relatively cool, dark place, preferably between 50 and 70°F
- Can no more food than you will use within a year.

Advantages of Hot Packing

Many fresh foods contain from 10 percent to more than 30 percent air. The length of time that food will last at premium quality depends on how much air is removed from the food before jars are sealed. The more air

that is removed, the higher the quality of the canned product.

Raw packing is the practice of filling jars tightly with freshly prepared but unheated food. Such foods, especially fruit, will float in the jars. The entrapped air in and around the food may cause discoloration within two to three months of storage. Raw packing is more suitable for vegetables processed in a pressure canner.

Hot packing is the practice of heating freshly prepared food to boiling, simmering it three to five minutes, and promptly filling jars loosely with the boiled food. Hot packing is the best way to remove air and is the preferred pack style for foods processed in a boiling-water canner. At first, the color of hot-packed foods may appear no better than that of raw-packed foods, but within a short storage period, both color and flavor of hot-packed foods will be superior.

Whether food has been hot packed or raw packed, the juice, syrup, or water to be added to the foods should be heated to boiling before it is added to the jars. This practice helps to remove air from food tissues, shrinks food, helps keep the food from floating in the jars, increases vacuum in sealed jars, and improves shelf life. Preshrinking food allows you to add more food to each jar.

Controlling Headspace

The unfilled space above the food in a jar and below its lid is termed headspace. It is best to leave a ¼-inch headspace for jams and jellies, ½-inch for fruits and tomatoes to be processed in boiling water, and from 1 to 1¼ inches in low-acid foods to be processed in a pressure canner.

This space is needed for expansion of food as jars are processed and for forming vacuums in cooled jars. The extent of expansion is determined by the air content in the food and by the processing temperature. Air expands greatly when heated to high temperatures—the higher the temperature, the greater the expansion. Foods expand less than air when heated.

Jars and Lids

Food may be canned in glass jars or metal containers. Metal containers can be used only once. They require special sealing equipment and are much more costly than jars.

Mason-type jars designed for home canning are ideal for preserving food by pressure or boiling-water canning. Regular and wide-mouthed threaded mason jars with self-sealing lids are the best choices. They are available in half-pint, pint, 1½-pint, and quart sizes. The standard jar mouth opening is about 2⅜ inches. Wide-mouthed jars have openings of about 3 inches, making them more easily filled and emptied. Regular-mouth decorative jelly jars are available in 8-ounce and 12-ounce sizes.

With careful use and handling, mason jars may be reused many times, requiring only new lids each time. When lids are used properly, jar seals and vacuums are excellent.

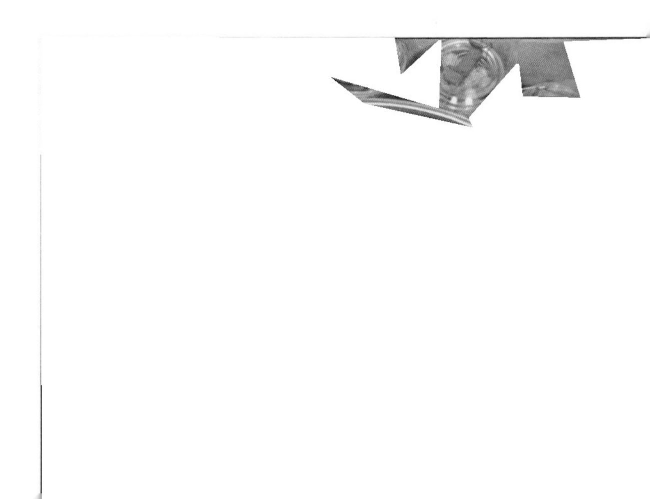

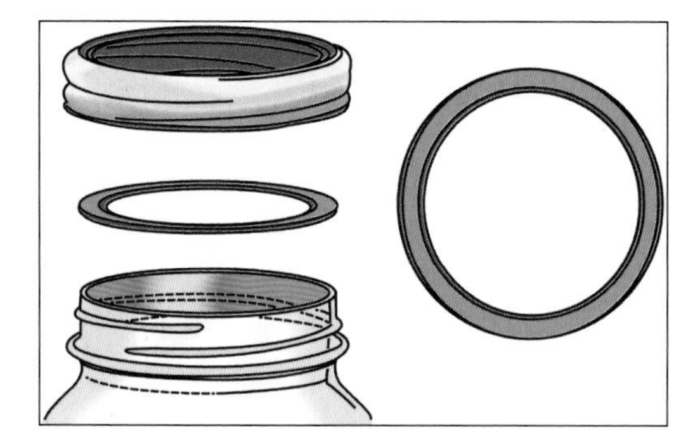

Jar Cleaning

Before reuse, wash empty jars in hot water with detergent and rinse well by hand, or wash in a dishwasher. Rinse thoroughly, as detergent residue may cause unnatural flavors and colors. Scale or hard-water films on jars are easily removed by soaking jars several hours in a solution containing 1 cup of vinegar (5 percent acid) per gallon of water.

Sterilization of Empty Jars

Use sterile jars for all jams, jellies, and pickled products processed less than ten minutes. To sterilize empty jars, put them right side up on the rack in a boiling-water canner. Fill the canner and jars with hot (not boiling) water to 1 inch above the tops of the jars. Boil ten minutes. Remove and drain hot sterilized jars one at a time. Save the hot water for processing filled jars. Fill jars with food, add lids, and tighten screw bands.

Empty jars used for vegetables, meats, and fruits to be processed in a pressure canner need not be sterilized beforehand. It is also unnecessary to sterilize jars for fruits, tomatoes, and pickled or fermented foods that will be processed ten minutes or longer in a boiling-water canner.

Lid Selection, Preparation, and Use

The common self-sealing lid consists of a flat metal lid held in place by a metal screw band during processing. The flat lid is crimped around its bottom edge to form a trough, which is filled with a colored gasket material. When jars are processed, the lid gasket softens and flows slightly to cover the jar-sealing surface, yet allows air to escape from the jar. The gasket then forms an airtight seal as the jar cools. Gaskets in unused lids work well for at least five years from date of manufacture. The gasket material in older unused lids may fail to seal on jars.

It is best to buy only the quantity of lids you will use in a year. To ensure a good seal, carefully follow the manufacturer's directions in preparing lids for use. Examine all metal lids carefully. Do not use old, dented, or deformed lids or lids with gaps or other defects in the sealing gasket.

After filling jars with food, release air bubbles by inserting a flat plastic (not metal) spatula between the food and the jar. Slowly turn the jar and move the spatula up and down to allow air bubbles to escape. Adjust the headspace and then clean the jar rim (sealing surface) with a dampened paper towel. Place the lid, gasket down, onto the cleaned jar-sealing surface. Uncleaned jar-sealing surfaces may cause seal failures.

Then fit the metal screw band over the flat lid. Follow the manufacturer's guidelines enclosed with or on the box for tightening the jar lids properly.

- If screw bands are too tight, air cannot vent during processing, and food will discolor during storage. Overtightening also may cause lids to buckle and jars to break, especially with raw-packed, pressure-processed food.
- If screw bands are too loose, liquid may escape from jars during processing, seals may fail, and the food will need to be reprocessed.

Do not retighten lids after processing jars. As jars cool, the contents in the jar contract, pulling the self-sealing lid firmly against the jar to form a high vacuum. Screw bands are not needed on stored jars. They can be removed easily after jars are cooled. When removed, washed, dried, and stored in a dry area, screw bands may be used many times.

If left on stored jars, they become difficult to remove, often rust, and may not work properly again.

Selecting the Correct Processing Time

When food is canned in boiling water, more processing time is needed for most raw-packed foods and for quart jars than is needed for hot-packed foods and pint jars.

To destroy microorganisms in acid foods processed in a boiling-water canner, you must:

- Process jars for the correct number of minutes in boiling water, and
- Cool the jars at room temperature.

To destroy microorganisms in low-acid foods processed with a pressure canner, you must:

- Process the jars for the correct number of minutes at 240°F (10 PSIG) or 250°F (15 PSIG), and
- Allow canner to cool at room temperature until it is completely depressurized.

The food may spoil if you fail to use the proper processing times, fail to vent steam from canners properly, process at lower pressure than specified, process for fewer minutes than specified, or cool the canner with water.

Processing times for haft-pint and pint jars are the same, as are times for 1 ½-pint and quart jars. For some products, you have a choice of processing at 5, 10, or 15 PSIG. In these cases, choose the canner pressure (PSIG) you wish to use and match it with your pack style (raw or hot) and jar size to find the correct processing time.

Recommended Canners

There are two main types of canners for heat-processing home-canned food: boiling-water canners and pressure canners. Most are designed to hold seven 1-quart jars or eight to nine 1-pint jars. Small pressure canners hold four one-quart jars; some large pressure canners hold eighteen 1-pint jars in two layers but hold only seven quart jars. Pressure saucepans with smaller volume capacities are not recommended for use in canning. Treat small pressure canners the same as standard larger canners; they should be vented using the typical venting procedures.

Low-acid foods must be processed in a pressure canner to be free of botulism risks. Although pressure canners also may be used for processing acid foods, boiling-water can-

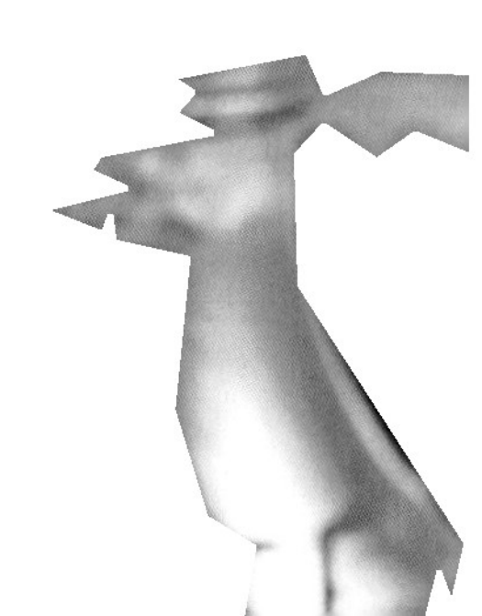

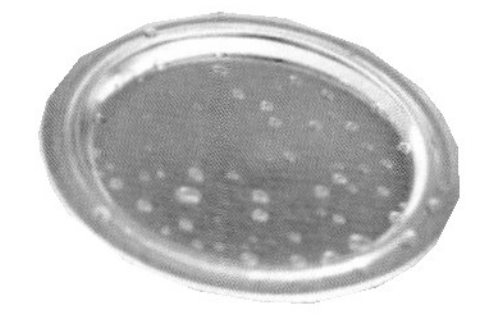

ners are recommended because they are faster. A pressure canner would require from fifty-five to 100 minutes to can a load of jars; the total time for canning most acid foods in boiling water varies from twenty-five to sixty minutes.

A boiling-water canner loaded with filled jars requires about twenty to thirty minutes of heating before its water begins to boil. A loaded pressure canner requires about twelve to fifteen minutes of heating before it begins to vent, another ten minutes to vent the canner, another five minutes to pressurize the canner, another eight to ten minutes to process the acid food, and, finally, another twenty to sixty minutes to cool the canner before removing jars.

Boiling-Water Canners

These canners are made of aluminum or porcelain-covered steel. They have removable perforated racks and fitted lids. The canner must be deep enough so that at least 1 inch of briskly boiling water will cover the tops of jars during processing. Some boiling-water canners do not have flat bottoms. A flat bottom must be used on an electric range. Either a flat or ridged bottom can be used on a gas burner. To ensure uniform processing of all jars with an electric range, the canner should be no more than 4 inches wider in diameter than the element on which it is heated.

Using a Boiling-Water Canner

Follow these steps for successful boiling-water canning:

1. Fill the canner halfway with water.
2. Preheat water to 140°F for raw-packed foods and to 180°F for hot-packed foods.

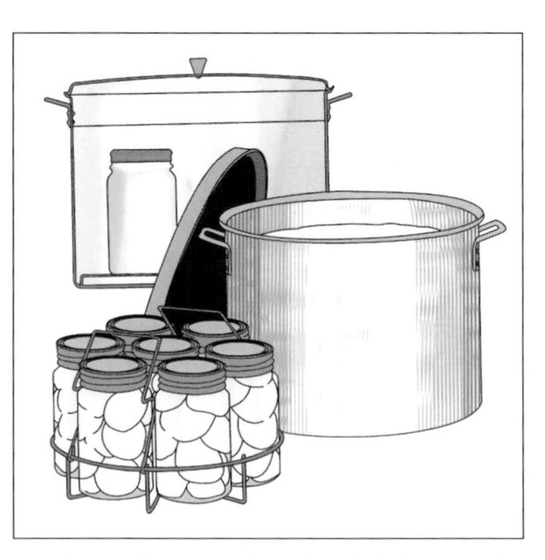

3. Load filled jars, fitted with lids, into the canner rack and use the handles to lower the rack into the water; or fill the canner, one jar at a time, with a jar lifter.
4. Add more boiling water, if needed, so the water level is at least 1 inch above jar tops.
5. Turn heat to its highest position until water boils vigorously.
6. Set a timer for the minutes required for processing the food.
7. Cover with the canner lid and lower the heat setting to maintain a gentle boil throughout the processing time.
8. Add more boiling water, if needed, to keep the water level above the jars.
9. When jars have been boiled for the recommended time, turn off the heat and remove the canner lid.
10. Using a jar lifter, remove the jars and place them on a towel, leaving at least 1 inch of space between the jars during cooling.

Pressure Canners

Pressure canners for use in the home have been extensively redesigned in recent years. Models made before the 1970s were heavy-walled kettles with clamp-on lids. They were fitted with a dial gauge, a vent port in the form of a petcock or counterweight, and a safety fuse. Modern pressure canners are lightweight, thin-walled kettles; most have

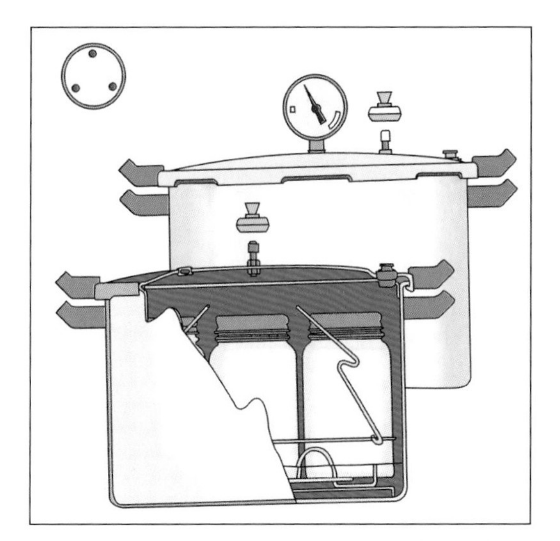

turn-on lids. They have a jar rack, gasket, dial or weighted gauge, an automatic vent or cover lock, a vent port (steam vent) that is closed with a counterweight or weighted gauge, and a safety fuse.

Pressure does not destroy microorganisms, but high temperatures applied for a certain period of time do. The success of destroying all microorganisms capable of growing in canned food is based on the temperature obtained in pure steam, free of air, at sea level. At sea level, a canner operated at a gauge pressure of 10 pounds provides an internal temperature of 240°F.

Air trapped in a canner lowers the inside temperature and results in under-processing. The highest volume of air trapped in a canner occurs in processing raw-packed foods in dial-gauge canners. These canners do not vent air during processing. To be safe, all types of pressure canners must be vented ten minutes before they are pressurized.

To vent a canner, leave the vent port uncovered on newer models or manually open petcocks on some older models. Heating the filled canner with its lid locked into place boils water and generates steam that escapes through the petcock or vent port. When steam first escapes, set a timer for ten minutes. After venting ten minutes, close the petcock or place the counterweight or weighted gauge over the vent port to pressurize the canner.

Weighted-gauge models exhaust tiny amounts of air and steam each time their gauge rocks or jiggles during processing. The sound of the weight rocking or jiggling indicates that the canner is maintaining the recommended pressure and needs no further attention until the load has been processed for the set time. Weighted-gauge canners cannot correct precisely for higher altitudes, and at altitudes above 1,000 feet must be operated at a pressure of 15.

Check dial gauges for accuracy before use each year and replace if they read high by more than 1 pound at 5, 10, or 15 pounds

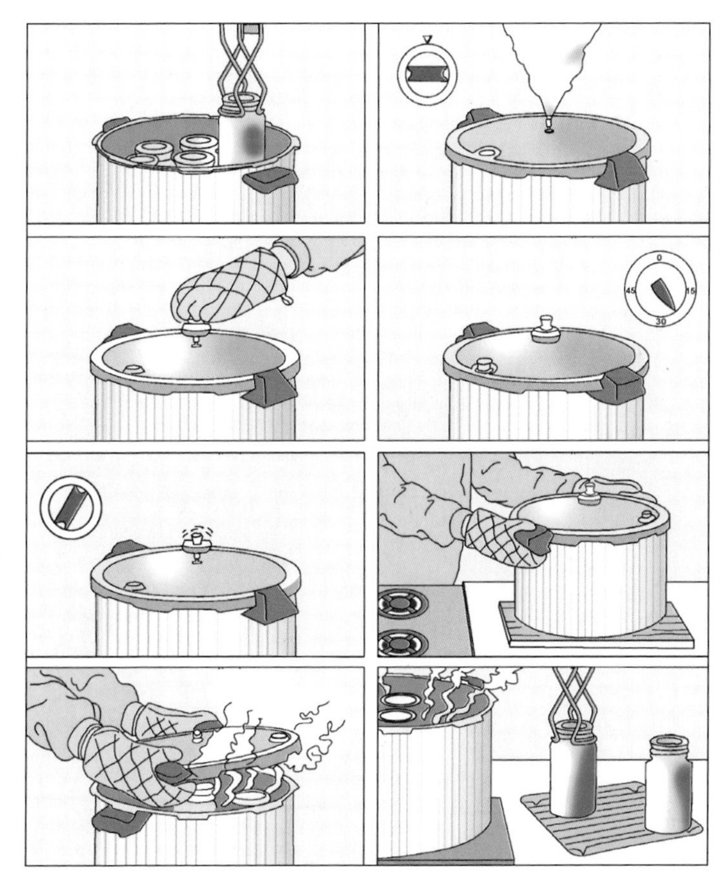

of pressure. Low readings cause over-processing and may indicate that the accuracy of the gauge is unpredictable. If a gauge is consistently low, you may adjust the processing pressure. For example, if the directions call for 12 pounds of pressure and your dial gauge has tested 1 pound low, you can safely process at 11 pounds of pressure. If the gauge is more than 2 pounds low, it is unpredictable, and it is best to replace it. Gauges may be checked at most USDA county extension offices, which are located in every state across the country. To find one near you, visit www.csrees.usda.gov.

Handle gaskets of canner lids carefully and clean them according to the manufacturer's directions. Nicked or dried gaskets will allow steam leaks during pressurization of canners. Gaskets of older canners may need to be lightly coated with vegetable oil once per year, but newer models are pre-lubricated. Check your canner's instructions.

Lid safety fuses are thin metal inserts or rubber plugs designed to relieve excessive pressure from the canner. Do not pick at or scratch fuses while cleaning lids. Use only canners that have Underwriters Laboratories' (UL) approval to ensure their safety.

Replacement gauges and other parts for canners are often available at stores offering canner equipment or from canner manufacturers. To order parts, list canner model number and describe the parts needed.

Using a Pressure Canner

Follow these steps for successful pressure canning:

1. Put 2 to 3 inches of hot water in the canner. Place filled jars on the rack, using a jar lifter. Fasten canner lid securely.
2. Open petcock or leave weight off vent port. Heat at the highest setting until steam flows from the petcock or vent port.
3. Maintain high heat setting, exhaust steam ten minutes, and then place weight on vent port or close petcock. The canner will pressurize during the next three to five minutes.
4. Start timing the process when the pressure reading on the dial gauge indicates that the recommended pressure has been reached or when the weighted gauge begins to jiggle or rock.
5. Regulate heat under the canner to maintain a steady pressure at or slightly above the correct gauge pressure. Quick and large pressure variations during processing may cause unnecessary liquid losses from jars. Weighted gauges on Mirro canners should jiggle about two or three times per minute. On Presto canners, they should rock slowly throughout the process.

When processing time is completed, turn off the heat, remove the canner from heat if possible, and let the canner depressurize. Do not force-cool the canner. If you cool it with cold running water in a sink or open the vent port before the canner depressurizes by itself, liquid will spurt from jars, causing low liquid levels and jar seal failures. Force-cooling also may warp the canner lid of older model canners, causing steam leaks.

Depressurization of older models should be timed. Standard size heavy-walled canners require about thirty minutes when loaded with pints and forty-five minutes with quarts. Newer thin-walled canners cool more rapidly and are equipped with vent locks. These canners are depressurized when their vent lock piston drops to a normal position.

1. After the vent port or petcock has been open for two minutes, unfasten the lid and carefully remove it. Lift the lid away from you so that the steam does not burn your face.
2. Remove jars with a lifter, and place on towel or cooling rack, if desired.

Cooling Jars

Cool the jars at room temperature for twelve to twenty-four hours. Jars may be cooled on racks or towels to minimize heat damage to counters. The food level and liquid volume of raw-packed jars will be noticeably lower after cooling because air is exhausted during processing and food shrinks. If a jar loses excessive liquid during processing, do not open it to add more liquid. As long as the seal is good, the product is still usable.

Testing Jar Seals

After cooling jars for twelve to twenty-four hours, remove the screw bands and test seals with one of the following methods:

Method 1: Press the middle of the lid with a finger or thumb. If the lid springs up when you release your finger, the lid is unsealed and reprocessing will be necessary.

Method 2: Tap the lid with the bottom of a teaspoon. If it makes a dull sound, the lid is not sealed. If food is in contact with the underside of the lid, it will also cause a dull sound. If the jar lid is sealed correctly, it will make a ringing, high-pitched sound.

Method 3: Hold the jar at eye level and look across the lid. The lid should be concave (curved down slightly in the center). If center of the lid is either flat or bulging, it may not be sealed.

Testing jar seals

Reprocessing Unsealed Jars

If a jar fails to seal, remove the lid and check the jar-sealing surface for tiny nicks. If necessary, change the jar, add a new, properly prepared lid, and reprocess within twenty-four hours using the same processing time.

Another option is to adjust headspace in unsealed jars to 1½ inches and freeze jars and contents instead of reprocessing. However, make sure jars have straight sides. Freezing may crack jars with "shoulders."

Foods in single unsealed jars could be stored in the refrigerator and consumed within several days.

Storing Canned Foods

If lids are tightly vacuum-sealed on cooled jars, remove screw bands, wash the lid and jar to remove food residue, then rinse and dry jars. Label and date the jars and store them in a clean, cool, dark, dry place. Do not store jars at temperatures above 95°F or near hot pipes, a range, a furnace, in an uninsulated attic, or in direct sunlight. Under these conditions, food will lose quality in a few weeks or months and may spoil. Dampness may corrode metal lids, break seals, and allow recontamination and spoilage.

Accidental freezing of canned foods will not cause spoilage unless jars become unsealed and recontaminated. However, freezing and thawing may soften food. If jars must be stored where they may freeze, wrap them in newspapers, place them in heavy cartons, and cover them with more newspapers and blankets.

Identifying and Handling Spoiled Canned Food

Growth of spoilage bacteria and yeast produces gas, which pressurizes the food, swells lids, and breaks jar seals. As each stored jar is selected for use, examine its lid for tightness and vacuum. Lids with concave centers have good seals.

Next, while holding the jar upright at eye level, rotate the jar and examine its outside surface for streaks of dried food originating at the top of the jar. Look at the contents for rising air bubbles and unnatural color.

While opening the jar, smell for unnatural odors and look for spurting liquid and cotton-like mold growth (white, blue, black, or green) on the top food surface and underside of lid. Do not taste food from a stored jar you discover to have an unsealed lid or that otherwise shows signs of spoilage.

All suspect containers of spoiled low-acid foods should be treated as having produced botulinum toxin and should be handled carefully as follows:

- If the suspect glass jars are unsealed, open, or leaking, they should be detoxified before disposal.
- If the suspect glass jars are sealed, remove lids and detoxify the entire jar, contents, and lids.

Detoxification Process

Carefully place the suspect containers and lids on their sides in an eight-quart-volume or larger stockpot, pan, or boiling-water canner. Wash your hands thoroughly. Carefully add water to the pot. The water should completely cover the containers with a minimum of 1 inch of water above the containers. Avoid splashing the water. Place a lid on the pot and heat the water to boiling. Boil thirty minutes to ensure detoxifying the food and all container components. Cool and discard lids and food in the trash or bury in the soil.

Thoroughly clean all counters, containers, and equipment including can opener, clothing, and hands that may have come in contact with the food or the containers. Discard any sponges or washcloths that were used in the cleanup. Place them in a plastic bag and discard in the trash.

Canned Foods for Special Diets

The cost of commercially canned special diet food often prompts interest in preparing these products at home. Some low-sugar and low-salt foods may be easily and safely canned at home. However, it may take some experimentation to create a product with the desired color, flavor, and texture. Start with a small batch and then make appropriate adjustments before producing large quantities.

How much should you can?

The amount of food to preserve for your family, either by canning or freezing, should be based on individual choices.

Fruit

There's nothing quite like opening a jar of home-preserved strawberries in the middle of a winter snowstorm. It takes you right back to the warm early-summer sunshine, the smell of the strawberry patch's damp earth, and the feel of the firm berries as you snipped them from the vines. Best of all, you get to indulge in the sweet, summery flavor even as the snow swirls outside the windows.

Preserving fruit is simple, safe, and it allows you to enjoy the fruits of your summer's labor all year round. On the next pages, you will find reference charts for processing various fruits and fruit products in a dial-gauge pressure canner or a weighted-gauge pressure canner. The same information is also included with each recipe's directions. In some cases, a boiling-water canner will serve better; for these instances, directions for its use are offered instead.

Adding syrup to canned fruit helps to retain its flavor, color, and shape, although it does not prevent spoilage. To maintain the most natural flavor, use the Very Light Syrup listed in the table found on page 98. Many fruits that are typically packed in heavy syrup are just as good—and a lot better for you—when packed in lighter syrups. However, if you're preserving fruit that's on the sour side, like cherries or tart apples, you might want to splurge on one of the sweeter versions.

Syrups

Adding syrup to canned fruit helps to retain its flavor, color, and shape, although jars still need to be processed to prevent spoilage. Follow the chart on page 98 for syrups of varying sweetness. Light corn syrups or mild-flavored honey may be used to replace up to half the table sugar called for in syrups.

> For hot packs, bring water and sugar to a boil, add fruit, reheat to a boil, and fill into jars immediately.

Directions

1. Bring water and sugar to a boil in a medium saucepan.
2. Pour over raw fruits in jars.

Apple Juice

The best apple juice is made from a blend of varieties. If you don't have your own apple press, try to buy fresh juice from a local cider maker within twenty-four hours after it has been pressed.

PROCESS TIMES FOR FRUITS AND FRUIT PRODUCTS IN A DIAL-GAUGE PRESSURE CANNER*

				Canner Pressure (PSI) at Altitudes of:			
Type of Fruit	Style of Pack	Jar Size	Process Time	0–2,000 ft	2,001–4,000 ft	4,001–6,000 ft	6,001–8,000 ft
Applesauce	Hot	Pints	8 minutes	6 lbs	7 lbs	8 lbs	9 lbs
	Hot	Quarts	10 minutes	6 lbs	7 lbs	8 lbs	9 lbs
Apples, sliced	Hot	Pints or Quarts	8 minutes	6 lbs	7 lbs	8 lbs	9 lbs
Berries, whole	Hot	Pints or Quarts	8 minutes	6 lbs	7 lbs	8 lbs	9 lbs
	Raw	Pints	8 minutes	6 lbs	7 lbs	8 lbs	9 lbs
	Raw	Quarts	10 minutes	6 lbs	7 lbs	8 lbs	9 lbs
Cherries, sour or sweet	Hot	Pints	8 minutes	6 lbs	7 lbs	8 lbs	9 lbs
	Hot	Quarts	10 minutes	6 lbs	7 lbs	8 lbs	9 lbs
	Raw	Pints or Quarts	10 minutes	6 lbs	7 lbs	8 lbs	9 lbs
Fruit purées	Hot	Pints or Quarts	8 minutes	6 lbs	7 lbs	8 lbs	9 lbs
Grapefruit or orange sections	Hot	Pints or Quarts	8 minutes	6 lbs	7 lbs	8 lbs	9 lbs
	Raw	Pints	8 minutes	6 lbs	7 lbs	8 lbs	9 lbs
	Raw	Quarts	10 minutes	6 lbs	7 lbs	8 lbs	9 lbs
Peaches, apricots, or nectarines	Hot or Raw	Pints or Quarts	10 minutes	6 lbs	7 lbs	8 lbs	9 lbs
Pears	Hot	Pints or Quarts	10 minutes	6 lbs	7 lbs	8 lbs	9 lbs
Plums	Hot or Raw	Pints or Quarts	10 minutes	6 lbs	7 lbs	8 lbs	9 lbs
Rhubarb	Hot	Pints or Quarts	8 minutes	6 lbs	7 lbs	8 lbs	9 lbs

*After the canner is completely depressurized, remove the weight from the vent port or open the petcock. Wait ten minutes; then unfasten the lid and remove it carefully. Lift the lid with the underside away from you so that the steam coming out of the canner does not burn your face.

PROCESS TIMES FOR FRUITS AND FRUIT PRODUCTS IN A WEIGHTED-GAUGE PRESSURE CANNER*

				Canner Pressure (PSI) at Altitudes of:	
Type of Fruit	Style of Pack	Jar Size	Process Time	0–1,000 ft	Above 1,000 ft
Applesauce	Hot	Pints	8 minutes	5 lbs	10 lbs
	Hot	Quarts	10 minutes	5 lbs	10 lbs
Apples, sliced	Hot	Pints or Quarts	8 minutes	5 lbs	10 lbs
Berries, whole	Hot	Pints or Quarts	8 minutes	5 lbs	10 lbs
	Raw	Pints	8 minutes	5 lbs	10 lbs
	Raw	Quarts	10 minutes	5 lbs	10 lbs
Cherries, sour or sweet	Hot	Pints	8 minutes	5 lbs	10 lbs
	Hot	Quarts	10 minutes	5 lbs	10 lbs
	Raw	Pints or Quarts	10 minutes	5 lbs	10 lbs
Fruit purées	Hot	Pints or Quarts	8 minutes	5 lbs	10 lbs
Grapefruit or orange sections	Hot	Pints or Quarts	8 minutes	5 lbs	10 lbs
	Raw	Pints	8 minutes	5 lbs	10 lbs
	Raw	Quarts	10 minutes	5 lbs	10 lbs
Peaches, apricots, or nectarines	Hot or Raw	Pints or Quarts	10 minutes	5 lbs	10 lbs
Pears	Hot	Pints or Quarts	10 minutes	5 lbs	10 lbs
Plums	Hot or Raw	Pints or Quarts	10 minutes	5 lbs	10 lbs
Rhubarb	Hot	Pints or Quarts	8 minutes	5 lbs	10 lbs

*After the canner is completely depressurized, remove the weight from the vent port or open the petcock. Wait ten minutes; then unfasten the lid and remove it carefully. Lift the lid with the underside away from you so that the steam coming out of the canner does not burn your face.

SUGAR AND WATER IN SYRUP

Syrup Type	Approx. % Sugar	Measures of Water and Sugar				Fruits Commonly Packed in Syrup
		For 9-Pt Load*		For 7-Qt Load		
		Cups Water	Cups Sugar	Cups Water	Cups Sugar	
Very Light	10	6½	¾	10½	1¼	Approximates natural sugar levels in most fruits and adds the fewest calories.
Light	20	5¾	1½	9	2¼	Very sweet fruit. Try a small amount the first time to see if your family likes it.
Medium	30	5¼	2¼	8¼	3¾	Sweet apples, sweet cherries, berries, grapes.
Heavy	40	5	3¼	7¾	5¼	Tart apples, apricots, sour cherries, gooseberries, nectarines, peaches, pears, plums.

Continued

Syrup Type	Approx. % Sugar	Measures of Water and Sugar — For 9-Pt Load*: Cups Water	For 9-Pt Load*: Cups Sugar	For 7-Qt Load: Cups Water	For 7-Qt Load: Cups Sugar	Fruits Commonly Packed in Syrup
Very Heavy	50	4¼	4¼	6½	6¾	Very sour fruit. Try a small amount the first time to see if your family likes it.

*This amount is also adequate for a four-quart load.

Canning Without Sugar

In canning regular fruits without sugar, it is very important to select fully ripe but firm fruits of the best quality. It is generally best to can fruit in its own juice, but blends of unsweetened apple, pineapple, and white grape juice are also good for pouring over solid fruit pieces. Adjust headspaces and lids and use the processing recommendations for regular fruits. Add sugar substitutes, if desired, when serving.

PROCESS TIMES FOR APPLE JUICE IN A BOILING-WATER CANNER*

Style of Pack	Jar Size	Process Time at Altitudes of: 0–1,000 ft	1,001–6,000 ft	Above 6,000 ft
Hot	Pints or Quarts	5 minutes	10 minutes	15 minutes
	Half-gallons	10 minutes	15 minutes	20 minutes

*After the process is complete, turn off the heat and remove the canner lid. Wait five minutes before removing jars.

PROCESS TIMES FOR APPLE BUTTER IN A BOILING-WATER CANNER*

Style of Pack	Jar Size	Process Time at Altitudes of: 0–1,000 ft	1,001–6,000 ft	Above 6,000 ft
Hot	Half-pints or Pints	5 minutes	10 minutes	15 minutes
	Quarts	10 minutes	15 minutes	20 minutes

*After the process is complete, turn off the heat and remove the canner lid. Wait five minutes before removing jars.

Directions

1. Refrigerate juice for twenty-four to forty-eight hours.
2. Without mixing, carefully pour off clear liquid and discard sediment. Strain the clear liquid through a paper coffee filter or double layers of damp cheesecloth.
3. Heat quickly in a saucepan, stirring occasionally, until juice begins to boil.
4. Fill immediately into sterile pint or quart jars or into clean half-gallon jars, leaving ¼-inch headspace.
5. Adjust lids and process. See chart above for recommended times for a boiling-water canner.

Apple Butter

The best apple varieties to use for apple butter include Jonathan, Winesap, Stayman, Golden Delicious, and Macintosh apples, but any of your favorite varieties will work. Don't bother to peel the apples, as you will strain the fruit before cooking it anyway. This recipe will yield eight to nine pints.

Ingredients

- 8 lbs. apples
- 2 cups cider
- 2 cups vinegar
- 2¼ cups white sugar
- 2¼ cups packed brown sugar
- 2 tbsp. ground cinnamon
- 1 tbsp. ground cloves

Directions

1. Wash, stem, quarter, and core apples.
2. Cook slowly in cider and vinegar until soft. Press fruit through a colander, food mill, or strainer.
3. Cook fruit pulp with sugar and spices, stirring frequently. To test for doneness, remove a spoonful and hold it away from steam for 2 minutes. If the butter remains mounded on the spoon, it is done. If you're still not sure, spoon a small quantity onto a plate. When a rim of liquid does not separate around the edge of the butter, it is ready for canning.
4. Fill hot into sterile half-pint or pint jars, leaving ¼-inch headspace. Quart jars need not be pre-sterilized.

Applesauce

Besides being delicious on its own or paired with dishes like pork chops or latkes, applesauce can be used as a butter substitute in many baked goods. Select apples that are sweet, juicy, and crisp. For a tart flavor, add one to two pounds of tart apples to each three pounds of sweeter fruit.

Directions

1. Wash, peel, and core apples. Slice apples into water containing a little lemon juice to prevent browning.
2. Place drained slices in an 8- to 10-quart pot. Add ½ cup water. Stirring occasionally to prevent burning, heat quickly until tender (5 to 20 minutes, depending on maturity and variety).
3. Press through a sieve or food mill, or skip the pressing step if you prefer chunky-style sauce. Sauce may be packed without sugar, but if desired, sweeten to taste (start with ⅛ cup sugar per quart of sauce).
4. Reheat sauce to boiling. Fill jars with hot sauce, leaving ½-inch headspace. Adjust lids and process.

Spiced Apple Rings

- 12 lbs. firm tart apples (maximum diameter 2-½ inches)
- 12 cups sugar
- 6 cups water
- 1¼ cups white vinegar (5%)

QUANTITY

1. An average of 21 pounds of apples is needed per canner load of seven quarts.
2. An average of 13½ pounds of apples is needed per canner load of nine pints.
3. A bushel weighs 48 pounds and yields 14 to 19 quarts of sauce—an average of three pounds per quart.

PROCESS TIMES FOR APPLESAUCE IN A BOILING-WATER CANNER*

		Process Time at Altitudes of:			
Style of Pack	Jar Size	0–1,000 ft	1,001–3,000 ft	3,001–6,000 ft	Above 6,000 ft
Hot	Pints	15 minutes	20 minutes	20 minutes	25 minutes
	Quarts	20 minutes	25 minutes	30 minutes	35 minutes

*After the process is complete, turn off the heat and remove the canner lid. Wait five minutes before removing jars.

PROCESS TIMES FOR APPLESAUCE IN A DIAL-GAUGE PRESSURE CANNER*

			Canner Pressure (PSI) at Altitudes of:			
Style of Pack	Jar Size	Process Time	0–2,000 ft	2,001–4,000 ft	4,001–6,000 ft	6,001–8,000 ft
Hot	Pints	8 minutes	6 lbs	7 lbs	8 lbs	9 lbs
	Quarts	10 minutes	6 lbs	7 lbs	8 lbs	9 lbs

*After the canner is completely depressurized, remove the weight from the vent port or open the petcock. Wait ten minutes; then unfasten the lid and remove it carefully. Lift the lid with the underside away from you so that the steam coming out of the canner does not burn your face.

PROCESS TIMES FOR APPLESAUCE IN A WEIGHTED-GAUGE PRESSURE CANNER*

			Canner Pressure (PSI) at Altitudes of:	
Style of Pack	Jar Size	Process Time	0–1,000 ft	Above 1,000 ft
Hot	Pints	8 minutes	5 lbs	10 lbs
	Quarts	10 minutes	5 lbs	10 lbs

*After the canner is completely depressurized, remove the weight from the vent port or open the petcock. Wait ten minutes; then unfasten the lid and remove it carefully. Lift the lid with the underside away from you so that the steam coming out of the canner does not burn your face.

- 3 tbsps whole cloves
- ¾ cup red hot cinnamon candies or 8 cinnamon sticks
- 1 tsp red food coloring (optional)

Yield: About 8 to 9 pints

Directions

1. Wash apples. To prevent discoloration, peel and slice one apple at a time. Immediately cut crosswise into ½-inch slices, remove core area with a melon baller, and immerse in ascorbic acid solution.
2. To make flavored syrup, combine sugar water, vinegar, cloves, cinnamon candies,

or cinnamon sticks and food coloring in a 6-quart saucepan. Stir, heat to boil, and simmer 3 minutes.

3. Drain apples, add to hot syrup, and cook 5 minutes. Fill jars (preferably wide-mouth) with apple rings and hot flavored syrup, leaving ½-inch headspace. Adjust lids and process according to the chart below.

Apricots, Halved or Sliced

Apricots are excellent in baked goods, stuffing, chutney, or on their own. Choose firm, well-colored mature fruit for best results.

Directions

1. Dip fruit in boiling water for 30 to 60 seconds until skins loosen. Dip quickly in cold water and slip off skins.

QUANTITY

- An average of 16 pounds of apricots is needed per canner load of 7 quarts.
- An average of 10 pounds is needed per canner load of 9 pints.
- A bushel weighs 50 pounds and yields 20 to 25 quarts—an average of 2¼ pounds per quart.

2. Cut in half, remove pits, and slice if desired. To prevent darkening, keep peeled fruit in water with a little lemon juice.
3. Prepare and boil a very light, light, or medium syrup (see page 98) or pack apricots in water, apple juice, or white grape juice.

Berries, Whole

Preserved berries are perfect for use in pies, muffins, pancakes, or in poultry or pork dressings. Nearly every berry preserves well, including blackberries, blueberries, currants, dewberries, elderberries, gooseberries, huckleberries, loganberries, mulberries, and raspberries. Choose ripe, sweet berries with uniform color.

Directions

1. Wash 1 or 2 quarts of berries at a time. Drain, cap, and stem if necessary. For gooseberries, snip off heads and tails with scissors.
2. Prepare and boil preferred syrup, if desired (see page 98). Add ½ cup syrup, juice, or water to each clean jar.

Hot pack—(Best for blueberries, currants, elderberries, gooseberries, and

(continued on page 104)

PROCESS TIME FOR SPICED APPLE RINGS IN A BOILING-WATER CANNER*

		Process Time at Altitudes of		
Style of Pack	Jar Size	0–1,000 ft	1,001–6,000 ft	Above 6,000 ft
Hot	Half-pints or Pints	10 minutes	15 minutes	20 minutes

*After the process is complete, turn off the heat and remove the canner lid. Wait five minutes before removing jars.

PROCESS TIMES FOR HALVED OR SLICED APRICOTS IN A DIAL-GAUGE PRESSURE CANNER*

			Canner Pressure (PSI) at Altitudes of:			
Style of Pack	**Jar Size**	**Process Time**	**0–2,000 ft**	**2,001–4,000 ft**	**4,001–6,000 ft**	**6,001–8,000 ft**
Hot or Raw	Pints or Quarts	10 minutes	6 lbs	7 lbs	8 lbs	9 lbs

*After the canner is completely depressurized, remove the weight from the vent port or open the petcock. Wait ten minutes; then unfasten the lid and remove it carefully. Lift the lid with the underside away from you so that the steam coming out of the canner does not burn your face.

PROCESS TIMES FOR HALVED OR SLICED APRICOTS IN A WEIGHTED-GAUGE PRESSURE CANNER*

			Canner Pressure (PSI) at Altitudes of:	
Style of Pack	**Jar Size**	**Process Time**	**0–1,000 ft**	**Above 1,000 ft**
Hot or Raw	Pints or Quarts	10 minutes	5 lbs	10 lbs

*After the canner is completely depressurized, remove the weight from the vent port or open the petcock. Wait ten minutes; then unfasten the lid and remove it carefully. Lift the lid with the underside away from you so that the steam coming out of the canner does not burn your face.

RECOMMENDED PROCESS TIMES FOR WHOLE BERRIES IN A BOILING-WATER CANNER*

		Process Time at Altitudes of:			
Style of Pack	**Jar Size**	**0–1,000 ft**	**1,001–3,000 ft**	**3,001–6,000 ft**	**Above 6,000 ft**
Hot	Pints or Quarts	15 minutes	20 minutes	20 minutes	25 minutes
Raw	Pints	15 minutes	20 minutes	20 minutes	25 minutes
	Quarts	20 minutes	25 minutes	30 minutes	35 minutes

*After the process is complete, turn off the heat and remove the canner lid. Wait five minutes before removing jars.

PROCESS TIMES FOR WHOLE BERRIES IN A DIAL-GAUGE PRESSURE CANNER*

			Canner Pressure (PSI) at Altitudes of:			
Style of Pack	**Jar Size**	**Process Time**	**0–2,000 ft**	**2,001–4,000 ft**	**4,001–6,000 ft**	**6,001–8,000 ft**
Hot	Pints or Quarts	8 minutes	6 lbs	7 lbs	8 lbs	9 lbs
Raw	Pints	8 minutes	6 lbs	7 lbs	8 lbs	9 lbs
Raw	Quarts	10 minutes	6 lbs	7 lbs	8 lbs	9 lbs

*After the canner is completely depressurized, remove the weight from the vent port or open the petcock. Wait ten minutes; then unfasten the lid and remove it carefully. Lift the lid with the underside away from you so that the steam coming out of the canner does not burn your face.

QUANTITY

- An average of 12 pounds of berries is needed per canner load of seven quarts.
- An average of 7 pounds is needed per canner load of 9 pints.
- A 24-quart crate weighs 36 pounds and yields 18 to 24 quarts—an average of 1¾ pounds per quart.

PROCESS TIMES FOR WHOLE BERRIES IN A WEIGHTED-GAUGE PRESSURE CANNER*

			Canner Pressure (PSI) at Altitudes of:	
Style of Pack	Jar Size	Process Time	0–1,000 ft	Above 1,000 ft
Hot	Pints or Quarts	8 minutes	5 lbs	10 lbs
Raw	Pints	8 minutes	5 lbs	10 lbs
Raw	Quarts	10 minutes	5 lbs	10 lbs

*After the canner is completely depressurized, remove the weight from the vent port or open the petcock. Wait ten minutes; then unfasten the lid and remove it carefully. Lift the lid with the underside away from you so that the steam coming out of the canner does not burn your face.

PROCESS TIMES FOR BERRY SYRUP IN A BOILING-WATER CANNER*

		Process Time at Altitudes of:		
Style of Pack	Jar Size	0–1,000 ft	1,001–6,000 ft	Above 6,000 ft
Hot	Half-pints or Pints	10 minutes	15 minutes	20 minutes

*After the process is complete, turn off the heat and remove the canner lid. Wait five minutes before removing jars.

To make syrup with whole berries, rather than crushed, save 1 or 2 cups of the fresh or frozen fruit, combine these with the sugar, and simmer until soft. Remove from heat, skim off foam, and fill into clean jars, following processing directions for regular berry syrup.

huckleberries) Heat berries in boiling water for 30 seconds and drain. Fill jars and cover with hot juice, leaving ½-inch headspace.

Raw pack—Fill jars with any of the raw berries, shaking down gently while filling. Cover with hot syrup, juice, or water, leaving ½-inch headspace.

Berry Syrup

Juices from fresh or frozen blueberries, cherries, grapes, raspberries (black or red), and strawberries are easily made into toppings for use on ice cream and pastries. For an elegant finish to cheesecakes or pound cakes, drizzle a thin stream in a zigzag across the top just before serving. Berry syrups are also great additions to smoothies or milkshakes. This recipe makes about 9 half-pints.

Directions

1. Select 6½ cups of fresh or frozen berries of your choice. Wash, cap, and stem berries and crush in a saucepan.
2. Heat to boiling and simmer until soft (5 to 10 minutes). Strain hot through a colander placed in a large pan and drain until cool enough to handle.
3. Strain the collected juice through a double layer of cheesecloth or jelly bag. Discard the dry pulp. The yield of the pressed juice should be about 4½ to 5 cups.

4. Combine the juice with 6¾ cups of sugar in a large saucepan, bring to a boil, and simmer 1 minute.
5. Fill into clean half-pint or pint jars, leaving ½-inch headspace. Adjust lids and process.

Fruit Purées

Almost any fruit can be puréed for use as baby food, in sauces, or just as a nutritious snack. Puréed prunes and apples can be used as a butter replacement in many baked goods. Use this recipe for any fruit except figs and tomatoes.

Directions

1. Stem, wash, drain, peel, and remove pits if necessary. Measure fruit into large saucepan, crushing slightly if desired.
2. Add 1 cup hot water for each quart of fruit. Cook slowly until fruit is soft, stirring frequently. Press through sieve or food mill. If desired, add sugar to taste.
3. Reheat pulp to boil, or until sugar dissolves (if added). Fill hot into clean jars, leaving ¼-inch headspace. Adjust lids and process.

Grape Juice

Purple grapes are full of antioxidants and help to reduce the risk of heart disease, cancer, and Alzheimer's disease. For juice, select sweet, well-colored, firm, mature fruit.

Directions

1. Wash and stem grapes. Place grapes in a saucepan and add boiling water to cover. Heat and simmer slowly until skin is soft.

PROCESS TIMES FOR FRUIT PURÉES IN A BOILING-WATER CANNER*

		Process Time at Altitudes of:		
Style of Pack	Jar Size	0–1,000 ft	1,001–6,000 ft	Above 6,000 ft
Hot	Pints or Quarts	15 minutes	20 minutes	25 minutes

*After the process is complete, turn off the heat and remove the canner lid. Wait five minutes before removing jars.

PROCESS TIMES FOR FRUIT PURÉES IN A DIAL-GAUGE PRESSURE CANNER*

			Canner Pressure (PSI) at Altitudes of:			
Style of Pack	Jar Size	Process Time	0–2,000 ft	2,001–4,000 ft	4,001–6,000 ft	6,001–8,000 ft
Hot	Pints or Quarts	8 minutes	6 lbs	7 lbs	8 lbs	9 lbs

*After the canner is completely depressurized, remove the weight from the vent port or open the petcock. Wait ten minutes; then unfasten the lid and remove it carefully. Lift the lid with the underside away from you so that the steam coming out of the canner does not burn your face.

PROCESS TIMES FOR FRUIT PURÉES IN A WEIGHTED-GAUGE PRESSURE CANNER*

			Canner Pressure (PSI) at Altitudes of:	
Style of Pack	Jar Size	Process Time	0–1,000 ft	Above 1,000 ft
Hot	Pints or Quarts	8 minutes	5 lbs	10 lbs

*After the canner is completely depressurized, remove the weight from the vent port or open the petcock. Wait ten minutes; then unfasten the lid and remove it carefully. Lift the lid with the underside away from you so that the steam coming out of the canner does not burn your face.

2. Strain through a damp jelly bag or double layers of cheesecloth, and discard solids. Refrigerate juice for 24 to 48 hours.
3. Without mixing, carefully pour off clear liquid and save; discard sediment. If desired, strain through a paper coffee filter for a clearer juice.
4. Add juice to a saucepan and sweeten to taste. Heat and stir until sugar is dissolved. Continue heating with occasional stirring until juice begins to boil. Fill into jars immediately, leaving ¼-inch headspace. Adjust lids and process.

Peaches, Halved or Sliced

Peaches are delicious in cobblers, crisps, and muffins, or grilled for a unique cake topping. Choose ripe, mature fruit with minimal bruising.

Directions

1. Dip fruit in boiling water for 30 to 60 seconds until skins loosen. Dip quickly in cold water and slip off skins. Cut in half, remove pits, and slice if desired. To prevent darkening, keep peeled fruit in ascorbic acid solution.
2. Prepare and boil a very light, light, or medium syrup or pack peaches in water, apple juice, or white grape juice. Raw packs make poor quality peaches.

Hot pack—In a large saucepan, place drained fruit in syrup, water, or juice and

PROCESS TIMES FOR GRAPE JUICE IN A BOILING-WATER CANNER*

		Process Time at Altitudes of:		
Style of Pack	Jar Size	0–1,000 ft	1,001–6,000 ft	Above 6,000 ft
Hot	Pints or Quarts	5 minutes	10 minutes	15 minutes
	Half-gallons	10 minutes	15 minutes	20 minutes

*After the process is complete, turn off the heat and remove the canner lid. Wait five minutes before removing jars.

QUANTITY

- An average of 24½ pounds of grapes needed per canner load of 7 quarts.
- An average of 16 pounds per canner load of 9 pints.
- A lug weighs 26 pounds and yields 7 to 9 quarts of juice—an average of 3½ pounds per quart.

bring to boil. Fill jars with hot fruit and cooking liquid, leaving ½-inch headspace. Place halves in layers, cut side down.

Raw pack—Fill jars with raw fruit, cut side down, and add hot water, juice, or syrup, leaving ½-inch headspace.

3. Adjust lids and process.

Pears, Halved

Choose ripe, mature fruit for best results. For a special treat, fill halved pears with a mixture of chopped dried apricots, pecans, brown sugar, and butter; bake or microwave until warm and serve with vanilla ice cream.

Directions

1. Wash and peel pears. Cut lengthwise in halves and remove core. A melon baller or metal measuring spoon works well for coring pears. To prevent discoloration, keep pears in water with a little lemon juice.
2. Prepare a very light, light, or medium syrup (see page 98) or use apple juice, white grape juice, or water. Raw packs make poor quality pears. Boil drained pears 5 minutes in syrup, juice, or water. Fill jars with hot fruit and cooking liquid, leaving ½-inch headspace. Adjust lids and process.

QUANTITY

- An average of 17½ pounds of peas is needed per canner load of 7 quarts.
- An average of 11 pounds is needed per canner load of 9 pints.
- A bushel weighs 50 pounds and yields 16 to 25 quarts—an average of 2½ pounds per quart.

QUANTITY

- An average of 17½ pounds of peaches is needed per canner load of 7 quarts.
- An average of 11 pounds is needed per canner load of 9 pints.
- A bushel weighs 48 pounds and yields 16 to 24 quarts—an average of 2½ pounds per quart.

PROCESS TIMES FOR HALVED OR SLICED PEACHES IN A BOILING-WATER CANNER*

		Process Time at Altitudes of:			
Style of Pack	Jar Size	0–1,000 ft	1,001–3,000 ft	3,001–6,000 ft	Above 6,000 ft
Hot	Pints	20 minutes	25 minutes	30 minutes	35 minutes
	Quarts	25 minutes	30 minutes	35 minutes	40 minutes
Raw	Pints	25 minutes	30 minutes	35 minutes	40 minutes
	Quarts	30 minutes	35 minutes	40 minutes	45 minutes

*After the process is complete, turn off the heat and remove the canner lid. Wait five minutes before removing jars.

PROCESS TIMES FOR HALVED OR SLICED PEACHES IN A DIAL-GAUGE PRESSURE CANNER*

			Canner Pressure (PSI) at Altitudes of:			
Style of Pack	Jar Size	Process Time	0–2,000 ft	2,001–4,000 ft	4,001–6,000 ft	6,001–8,000 ft
Hot or Raw	Pints or Quarts	10 minutes	6 lbs	7 lbs	8 lbs	9 lbs

*After the canner is completely depressurized, remove the weight from the vent port or open the petcock. Wait ten minutes; then unfasten the lid and remove it carefully. Lift the lid with the underside away from you so that the steam coming out of the canner does not burn your face.

PROCESS TIMES FOR HALVED OR SLICED PEACHES IN A WEIGHTED-GAUGE PRESSURE CANNER*

			Canner Pressure (PSI) at Altitudes of:	
Style of Pack	Jar Size	Process Time	0–1,000 ft	Above 1,000 ft
Hot or Raw	Pints or Quarts	10 minutes	5 lbs	10 lbs

*After the canner is completely depressurized, remove the weight from the vent port or open the petcock. Wait ten minutes; then unfasten the lid and remove it carefully. Lift the lid with the underside away from you so that the steam coming out of the canner does not burn your face.

PROCESS TIMES FOR HALVED PEARS IN A BOILING-WATER CANNER*

		Process Time at Altitudes of:			
Style of Pack	Jar Size	0–1,000 ft	1,001–3,000 ft	3,001–6,000 ft	Above 6,000 ft
Hot	Pints	20 minutes	25 minutes	30 minutes	35 minutes
	Quarts	25 minutes	30 minutes	35 minutes	40 minutes

*After the process is complete, turn off the heat and remove the canner lid. Wait five minutes before removing jars.

PROCESS TIMES FOR HALVED PEARS IN A DIAL-GAUGE PRESSURE CANNER*

			Canner Pressure (PSI) at Altitudes of:			
Style of Pack	Jar Size	Process Time	0–2,000 ft	2,001–4,000 ft	4,001–6,000 ft	6,001–8,000 ft
Hot	Pints or Quarts	10 minutes	6 lbs	7 lbs	8 lbs	9 lbs

*After the canner is completely depressurized, remove the weight from the vent port or open the petcock. Wait ten minutes; then unfasten the lid and remove it carefully. Lift the lid with the underside away from you so that the steam coming out of the canner does not burn your face.

PROCESS TIMES FOR HALVED PEARS IN A WEIGHTED-GAUGE PRESSURE CANNER*

			Canner Pressure (PSI) at Altitudes of:	
Style of Pack	Jar Size	Process Time	0–1,000 ft	Above 1,000 ft
Hot	Pints or Quarts	10 minutes	5 lbs	10 lbs

*After the canner is completely depressurized, remove the weight from the vent port or open the petcock. Wait ten minutes; then unfasten the lid and remove it carefully. Lift the lid with the underside away from you so that the steam coming out of the canner does not burn your face.

Rhubarb, Stewed

Rhubarb in the garden is a sure sign that spring has sprung and summer is well on its way. But why not enjoy rhubarb all year round? The brilliant red stalks make it as appropriate for a holiday table as for an early summer feast. Rhubarb is also delicious in crisps, cobblers, or served hot over ice cream. Select young, tender, well-colored stalks from the spring or, if available, late fall crop.

QUANTITY

- An average of 10½ pounds of rhubarb is needed per canner load of 7 quarts.
- An average of 7 pounds is needed per canner load of 9 pints.
- A lug weighs 28 pounds and yields 14 to 28 quarts—an average of 1½ pounds per quart.

Directions

1. Trim off leaves. Wash stalks and cut into ½-inch to 1-inch pieces.
2. Place rhubarb in a large saucepan, and add ½ cup sugar for each quart of fruit. Let stand until juice appears. Heat gently to boiling. Fill jars without delay, leaving ½-inch headspace. Adjust lids and process.

Canned Pie Fillings

Using a premade pie filling will cut your pie preparation time by more than half, but most commercially produced fillings are oozing with high fructose corn syrup and all manner of artificial coloring and flavoring. (Food coloring is not at all necessary, but if you're really concerned about how the inside of your pie will look, appropriate amounts are added to each recipe as an optional ingredient.) Making and preserving your own pie fillings means that you can use your own fresh ingredients and adjust the sweetness to your taste. Because some folks like their pies

PROCESS TIMES FOR STEWED RHUBARB IN A BOILING-WATER CANNER*

		Process Time at Altitudes of:		
Style of Pack	Jar Size	0–1,000 ft	1,001–6,000 ft	Above 6,000 ft
Hot	Pints or Quarts	15 minutes	20 minutes	25 minutes

*After the process is complete, turn off the heat and remove the canner lid. Wait five minutes before removing jars.

PROCESS TIMES FOR STEWED RHUBARB IN A DIAL-GAUGE PRESSURE CANNER*

			Canner Pressure (PSI) at Altitudes of			
Style of Pack	Jar Size	Process Time	0–2,000 ft	2,001–4,000 ft	4,001–6,000 ft	6,001–8,000 ft
Hot	Pints or Quarts	8 minutes	6 lbs	7 lbs	8 lbs	9 lbs

*After the canner is completely depressurized, remove the weight from the vent port or open the petcock. Wait ten minutes; then unfasten the lid and remove it carefully. Lift the lid with the underside away from you so that the steam coming out of the canner does not burn your face.

PROCESS TIMES FOR STEWED RHUBARB IN A WEIGHTED-GAUGE PRESSURE CANNER*

			Canner Pressure (PSI) at Altitudes of:	
Style of Pack	Jar Size	Process Time	0–1,000 ft	Above 1,000 ft
Hot	Pints or Quarts	8 minutes	5 lbs	10 lbs

*After the canner is completely depressurized, remove the weight from the vent port or open the petcock. Wait ten minutes; then unfasten the lid and remove it carefully. Lift the lid with the underside away from you so that the steam coming out of the canner does not burn your face.

rich and sweet and others prefer a natural tart flavor, you might want to first make a single quart, make a pie with it, and see how you like it. Then you can adjust the sugar and spices in the recipe to suit your personal preferences before making a large batch. Experiment with combining fruits or adding different spices, but the amount of lemon juice should not be altered, as it aids in controlling the safety and storage stability of the fillings.

These recipes use Clear Jel® (sometimes sold as Clear Jel A®), a chemically modified cornstarch that produces excellent sauce consistency even after fillings are canned and baked. By using Clear Jel®, you can lower the sugar content of your fillings without sacrificing safety, flavor, or texture. (Note: Instant Clear Jel® is not meant to be cooked and should not be used for these recipes. Sure-Gel® is a natural fruit pectin and is not a suitable substitute for Clear Jel®. Cornstarch, tapioca starch, or arrowroot starch can be used in place of Clear Jel®, but the finished product is likely to be runny.) One pound of Clear Jel® costs less than five dollars and is enough to make fillings for about fourteen pies. It will keep for at least a year if stored in a cool, dry place. Clear Jel® is increasingly available among canning and freezing supplies in some stores. Alternately, you can order it by the pound at any of the following online stores:

- www.barryfarm.com
- www.kitchenkrafts.com
- www.theingredientstore.com

Apple Pie Filling

Use firm, crisp apples, such as Stayman, Golden Delicious, or Rome varieties, for the best results. If apples lack tartness, use an additional ¼ cup of lemon juice for each six quarts of slices. Ingredients are included for a 1-quart (enough for one 8-inch pie) or a 7-quart recipe.

Ingredients

	1 Quart	7 Quarts
Blanched, sliced fresh apples	3½ cups	6 quarts
Granulated sugar	¾ cup + 2 tbsp	5½ cups
Clear Jel®	¼ cup	1½ cup
Cinnamon	½ tsp	1 tbsp
Cold water	½ cup	2½ cups
Apple juice	¾ cup	5 cups
Bottled lemon juice	2 tbsp	¾ cup
Nutmeg (optional)	⅛ tsp	1 tsp

Directions

1. Wash, peel, and core apples. Prepare slices ½-inch wide and place in water containing a little lemon juice to prevent browning.
2. For fresh fruit, place 6 cups at a time in 1 gallon of boiling water. Boil each batch 1 minute after the water returns to a boil. Drain, but keep heated fruit in a covered bowl or pot.
3. Combine sugar, Clear Jel®, and cinnamon in a large kettle with water and apple juice. Add nutmeg, if desired. Stir and cook on medium-high heat until mixture thickens and begins to bubble.
4. Add lemon juice and boil 1 minute, stirring constantly. Fold in drained apple slices immediately and fill jars with mixture without delay, leaving 1-inch headspace. Adjust lids and process immediately.

Blueberry Pie Filling

Select fresh, ripe, and firm blueberries. Unsweetened frozen blueberries may be used. If sugar has been added, rinse it off while fruit is still frozen. Thaw fruit, then collect, measure, and use juice from fruit to partially replace the water specified in the recipe. Ingredients are included for a 1-quart (enough for one 8-inch pie) or 7-quart recipe.

Ingredients

	1 Quart	7 Quarts
Fresh or thawed blueberries	3½ cups	6 quarts
Granulated sugar	¾ cup + 2 tbsps	6 cups
Clear Jel®	¼ cup + 1 tbsp	2¼ cups
Cold water	1 cup	7 cups
Bottled lemon juice	3½ cups	½ cup
Blue food coloring (optional)	3 drops	20 drops
Red food coloring (optional)	1 drop	7 drops

When using frozen cherries and blueberries, select unsweetened fruit. If sugar has been added, rinse it off while fruit is frozen. Thaw fruit, then collect, measure, and use juice from fruit to partially replace the water specified in the recipe.

Directions

1. Wash and drain blueberries. Place 6 cups at a time in 1 gallon boiling water. Allow water to return to a boil and cook each batch for 1 minute. Drain but keep heated fruit in a covered bowl or pot.
2. Combine sugar and Clear Jel® in a large kettle. Stir. Add water and food coloring, if desired. Cook on medium-high heat until mixture thickens and begins to bubble.
3. Add lemon juice and boil 1 minute, stirring constantly. Fold in drained berries immediately and fill jars with mixture without delay, leaving 1-inch headspace. Adjust lids and process immediately.

Cherry Pie Filling

Select fresh, very ripe, and firm cherries. Unsweetened frozen cherries may be used. If sugar has been added, rinse it off while the fruit is still frozen. Thaw fruit, then collect, measure, and use juice from fruit to partially replace the water specified in the recipe. Ingredients are included for a 1-quart (enough for one 8-inch pie) or 7-quart recipe.

Ingredients

	1 Quart	7 Quarts
Fresh or thawed sour cherries	3⅓ cups	6 quarts
Granulated sugar	1 cup	7 cups
Clear Jel®	¼ cup + 1 tbsp	1¾ cups
Cold water	1⅓ cups	9⅓ cups
Bottled lemon juice	1 tbsp + 1 tsp	½ cup
Cinnamon (optional)	⅛ tsp	1 tsp
Almond extract (optional)	¼ tsp	2 tsps
Red food coloring (optional)	6 drops	¼ tsp

Directions

1. Rinse and pit fresh cherries, and hold in cold water. To prevent stem end brown-

PROCESS TIMES FOR APPLE PIE FILLING IN A BOILING-WATER CANNER*

		Process Time at Altitudes of:			
Style of Pack	Jar Size	0–1,000 ft	1,001–3,000 ft	3,001–6,000 ft	Above 6,000 ft
Hot	Pints or Quarts	25 minutes	30 minutes	35 minutes	40 minutes

*After the process is complete, turn off the heat and remove the canner lid. Wait five minutes before removing jars.

PROCESS TIMES FOR BLUEBERRY PIE FILLING IN A BOILING-WATER CANNER*

		Process Time at Altitudes of:			
Style of Pack	Jar Size	0–1,000 ft	1,001–3,000 ft	3,001–6,000 ft	Above 6,000 ft
Hot	Pints or Quarts	30 minutes	35 minutes	40 minutes	45 minutes

*After the process is complete, turn off the heat and remove the canner lid. Wait five minutes before removing jars.

ing, use water with a little lemon juice. Place 6 cups at a time in 1 gallon boiling water. Boil each batch 1 minute after the water returns to a boil. Drain but keep heated fruit in a covered bowl or pot.

2. Combine sugar and Clear Jel® in a large saucepan and add water. If desired, add cinnamon, almond extract, and food coloring. Stir mixture and cook over medium-high heat until mixture thickens and begins to bubble.
3. Add lemon juice and boil 1 minute, stirring constantly. Fold in drained cherries immediately and fill jars with mixture without delay, leaving 1-inch headspace. Adjust lids and process immediately.

Festive Mincemeat Pie Filling

Mincemeat pie originated as "Christmas Pie" in the eleventh century, when the English crusaders returned from the Holy Land bearing oriental spices. They added three of these spices—cinnamon, cloves, and nutmeg—to their meat pies to represent the three gifts that the magi brought to the Christ child. Mincemeat pies are traditionally small and are perfect paired with a mug of hot buttered rum. Walnuts or pecans can be used in place of meat if preferred. This recipe yields about seven quarts.

Ingredients

- 2 cups finely chopped suet
- 4 lbs. ground beef or 4 lbs. ground venison and 1 lb. sausage
- 5 qts. chopped apples
- 2 lbs. dark seedless raisins
- 1 lb. white raisins
- 2 qts. apple cider
- 2 tbsps ground cinnamon
- 2 tsps ground nutmeg
- ½ tsp cloves
- 5 cups sugar
- 2 tbsps salt

Directions

1. Cook suet and meat in water to avoid browning. Peel, core, and quarter apples. Put suet, meat, and apples through food grinder using a medium blade.
2. Combine all ingredients in a large saucepan, and simmer 1 hour or until slightly thickened. Stir often.
3. Fill jars with mixture without delay, leaving 1-inch headspace. Adjust lids and process.

Jams, Jellies, and Other Fruit Spreads

Homemade jams and jellies have lots more flavor than store-bought, over-processed varieties. The combinations of fruits and spices are limitless, so have fun experimenting with these recipes. If you can bear to part with your creations when you're all done, they make wonderful gifts for any occasion.

Pectin is what makes jams and jellies thicken and gel. Many fruits, such as crab apples, citrus fruits, sour plums, currants, quinces, green apples, or Concord grapes, have plenty of their own natural pectin, so there's no need to add more pectin to your

PROCESS TIMES FOR CHERRY PIE FILLING IN A BOILING-WATER CANNER*

		Process Time at Altitudes of:			
Style of Pack	Jar Size	0–1,000 ft	1,001–3,000 ft	3,001–6,000 ft	Above 6,000 ft
Hot	Pints or Quarts	30 minutes	35 minutes	40 minutes	45 minutes

*After the process is complete, turn off the heat and remove the canner lid. Wait five minutes before removing jars.

PROCESS TIMES FOR FESTIVE MINCEMEAT PIE FILLING IN A DIAL-GAUGE PRESSURE CANNER*

			Canner Pressure (PSI) at Altitudes of:			
Style of Pack	Jar Size	Process Time	0–2,000 ft	2,001–4,000 ft	4,001–6,000 ft	6,000–8,000 ft
Hot	Quarts	90 minutes	11 lbs	12 lbs	13 lbs	14 lbs

*After the canner is completely depressurized, remove the weight from the vent port or open the petcock. Wait ten minutes; then unfasten the lid and remove it carefully. Lift the lid with the underside away from you so that the steam coming out of the canner does not burn your face.

PROCESS TIMES FOR FESTIVE MINCEMEAT PIE FILLING IN A WEIGHTED-GAUGE PRESSURE CANNER*

			Canner Pressure (PSI) at Altitudes of:	
Style of Pack	Jar Size	Process Time	0–1,000 ft	Above 1,000 ft
Hot	Quarts	90 minutes	10 lbs	15 lbs

*After the canner is completely depressurized, remove the weight from the vent port or open the petcock. Wait ten minutes; then unfasten the lid and remove it carefully. Lift the lid with the underside away from you so that the steam coming out of the canner does not burn your face.

TIP

If you are not sure if a fruit has enough of its own pectin, combine 1 tablespoon of rubbing alcohol with 1 tablespoon of extracted fruit juice in a small glass. Let stand 2 minutes. If the mixture forms into one solid mass, there's plenty of pectin. If you see several weak blobs, you need to add pectin or combine with another high-pectin fruit.

recipes. You can use less sugar when you don't add pectin, but you will have to boil the fruit for longer. Still, the process is relatively simple and you don't have to worry about having store-bought pectin on hand.

To use fresh fruits with a low pectin content or canned or frozen fruit juice, powdered or liquid pectin must be added for your jams and jellies to thicken and set properly. Jelly or jam made with added pectin requires less cooking and generally gives a larger yield. These products have more natural fruit flavors, too. In addition, using added pectin eliminates the need to test hot jellies and jams for proper gelling.

Beginning this section are descriptions of the differences between methods and tips for success with whichever you use.

Making Jams and Jellies without Added Pectin

Jelly without Added Pectin

Making jelly without added pectin is not an exact science. You can add a little more or less sugar according to your taste, substitute honey for up to half of the sugar, or experiment with combining small amounts of low-pectin fruits with other high-pectin fruits. The Ingredients table on page 116 shows you the basics for common high-pectin fruits. Use it as a guideline as you experiment with other fruits.

As fruit ripens, its pectin content decreases, so use fruit that has recently been picked, and mix ¾ ripe fruit with ¼ under-ripe. Cooking cores and peels along with the fruit will also increase the pectin level. Avoid using canned or frozen fruit as they contain very little pectin. Be sure to wash all fruit thoroughly before cooking. One pound of fruit should yield at least 1 cup of clear juice.

Directions

1. Crush soft fruits or berries; cut firmer fruits into small pieces (there is no need to peel or core the fruits, as cooking all the parts adds pectin).
2. Add water to fruits that require it, as listed in the Ingredients table on the next page. Put fruit and water in large saucepan and bring to a boil. Then simmer according to the times below until fruit is soft, while stirring to prevent scorching.
3. When fruit is tender, strain through a colander, then strain through a double layer of cheesecloth or a jelly bag. Allow juice to drip through, using a stand or colander to hold the bag. Avoid pressing or squeezing the bag or cloth as it will cause cloudy jelly.
4. Using no more than 6 to 8 cups of extracted fruit juice at a time, measure fruit juice, sugar, and lemon juice according to the Ingredients table, and heat to boiling.
5. Stir until the sugar is dissolved. Boil over high heat to the jellying point. To test jelly for doneness, follow the steps on the next page.
6. Remove from heat and quickly skim off foam. Fill sterile jars with jelly. Use a measuring cup or ladle the jelly through a wide-mouthed funnel, leaving ¼-inch headspace. Adjust lids and process.

TIP

Commercially frozen and canned juices may be low in natural pectins and make soft textured spreads.

Preventing spoilage

Even though sugar helps preserve jellies and jams, molds can grow on the surface of these products. Research now indicates that the mold which people usually scrape off the surface of jellies may not be as harmless as it seems. Mycotoxins have been found in some jars of jelly having surface mold growth. Mycotoxins are known to cause cancer in animals; their effects on humans are still being researched. Because of possible mold contamination, paraffin or wax seals are no longer recommended for any sweet spread, including jellies. To prevent growth of molds and loss of good flavor or color, fill products hot into sterile Mason jars, leaving ¼-inch headspace, seal with self-sealing lids, and process five minutes in a boiling-water canner. Correct process time at higher elevations by adding one additional minute per 1,000 feet above sea level. If unsterile jars are used, the filled jars should be processed ten minutes. Use of sterile jars is preferred, especially when fruits are low in pectin, since the added five-minute process time may cause weak gels.

Lemon Curd

Lemon curd is a rich, creamy spread that can be used on (or in) a variety of teatime treats—crumpets, scones, cake fillings, tartlets, or meringues are all enhanced by its tangy-sweet flavor. Follow the recipe carefully, as variances in ingredients, order, and temperatures may lead to a poor texture or flavor. For Lime Curd, use the same recipe but substitute 1 cup bottled lime juice and ¼ cup fresh lime zest for the lemon juice and zest. This recipe yields about 3 to 4 half-pints.

Ingredients

- 2½ cups superfine sugar*
- ½ cup lemon zest (freshly zested), optional
- 1 cup bottled lemon juice**

Ingredients

Fruit	Water to be Added per Pound of Fruit	Minutes to Simmer Fruit before Extracting Juice	Ingredients Added to Each Cup of Strained Juice		Yield from 4 Cups of Juice (Half-pints)
			Sugar (Cups)	Lemon Juice (Tsp)	
Apples	1 cup	20 to 25	¾	1½ (opt)	4 to 5
Blackberries	None or ¼ cup	5 to 10	¾ to 1	None	7 to 8
Crab apples	1 cup	20 to 25	1	None	4 to 5
Grapes	None or ¼ cup	5 to 10	¾ to 1	None	8 to 9
Plums	½ cup	15 to 20	¾	None	8 to 9

Temperature test—Use a jelly or candy thermometer and boil until mixture reaches the following temperatures:

Sea Level	1,000 ft	2,000 ft	3,000 ft	4,000 ft	5,000 ft	6,000 ft	7,000 ft	8,000 ft
220°F	218°F	216°F	214°F	212°F	211°F	209°F	207°F	205°F

Sheet or spoon test—Dip a cool metal spoon into the boiling jelly mixture. Raise the spoon about 12 inches above the pan (out of steam). Turn the spoon so the liquid runs off the side. The jelly is done when the syrup forms two drops that flow together and sheet or hang off the edge of the spoon.

PROCESS TIMES FOR JELLY WITHOUT ADDED PECTIN IN A BOILING WATER CANNER*

		Process Time at Altitudes of:		
Style of Pack	Jar Size	0–1,000 ft	1,001–6,000 ft	Above 6,000 ft
Hot	Half-pints or pints	5 minutes	10 minutes	15 minutes

*After the process is complete, turn off the heat and remove the canner lid. Wait five minutes before removing jars.

- ¾ cup unsalted butter, chilled, cut into approximately ¾-inch pieces
- 7 large egg yolks
- 4 large whole eggs

Directions

1. Wash 4 half-pint canning jars with warm, soapy water. Rinse well; keep hot until ready to fill. Prepare canning lids according to manufacturer's directions.
2. Fill boiling water canner with enough water to cover the filled jars by 1 to 2 inches. Use a thermometer to preheat the water to 180°F by the time filled jars are ready to be added. **Caution:** Do not heat the water in the canner to more than 180°F before jars are added. If the water in the canner is too hot when jars are added, the process time will not be long enough. The time it takes for the canner to reach boiling after the jars are added is expected to be 25 to 30 minutes for this product. Process time starts after the water in the canner comes to a full boil over the tops of the jars.
3. Combine the sugar and lemon zest in a small bowl, stir to mix, and set aside about 30 minutes. Premeasure the lemon juice and prepare the chilled butter pieces.
4. Heat water in the bottom pan of a double boiler*** until it boils gently. The water should not boil vigorously or touch the bottom of the top double boiler pan or bowl in which the curd is to be cooked. Steam produced will be sufficient for the cooking process to occur.
5. In the top of the double boiler, on the counter top or table, whisk the egg yolks and whole eggs together until thoroughly mixed. Slowly whisk in the sugar and zest, blending until well mixed and smooth. Blend in the lemon juice and then add the butter pieces to the mixture.
6. Place the top of the double boiler over boiling water in the bottom pan. Stir gently but continuously with a silicone spatula or cooking spoon, to prevent the mixture from sticking to the bottom of the pan. Continue cooking until the mixture reaches a temperature of 170°F. Use a food thermometer to monitor the temperature.
7. Remove the double boiler pan from the stove and place on a protected surface, such as a dishcloth or towel on the counter top. Continue to stir gently until the curd thickens (about 5 minutes). Strain curd through a mesh strainer into a glass or stainless steel bowl; discard collected zest.
8. Fill hot strained curd into the clean, hot half-pint jars, leaving ½-inch headspace. Remove air bubbles and adjust headspace if needed. Wipe rims of jars

---

* If superfine sugar is not available, run granulated sugar through a grinder or food processor for 1 minute, let settle, and use in place of superfine sugar. Do not use powdered sugar.

** Bottled lemon juice is used to standardize acidity. Fresh lemon juice can vary in acidity and is not recommended.

*** If a double boiler is not available, a substitute can be made with a large bowl or saucepan that can fit partway down into a saucepan of a smaller diameter. If the bottom pan has a larger diameter, the top bowl or pan should have a handle or handles that can rest on the rim of the lower pan.

PROCESS TIMES FOR LEMON CURD IN A BOILING-WATER CANNER*

		Process Time at Altitudes of:		
Style of Pack	Jar Size	0–1,000 ft	1,001–6,000 ft	Above 6,000 ft
Hot	Half-pints	15 minutes	20 minutes	25 minutes

*After the process is complete, turn off the heat and remove the canner lid. Wait five minutes before removing jars.

with a dampened, clean paper towel; apply two-piece metal canning lids. Process. Let cool, undisturbed, for twelve to twenty-four hours and check for seals.

Jam without Added Pectin

Making jam is even easier than making jelly, as you don't have to strain the fruit. However, you'll want to be sure to remove all stems, skins, and pits. Be sure to wash and rinse all fruits thoroughly before cooking, but don't let them soak. For best flavor, use fully ripe fruit. Use the ingredients table below as a guideline as you experiment with less common fruits.

1. Remove stems, skins, seeds, and pits; cut into pieces and crush. For berries, remove stems and blossoms and crush. Seedy berries may be put through a sieve or food mill. Measure crushed fruit into large saucepan using the ingredient quantities specified above.

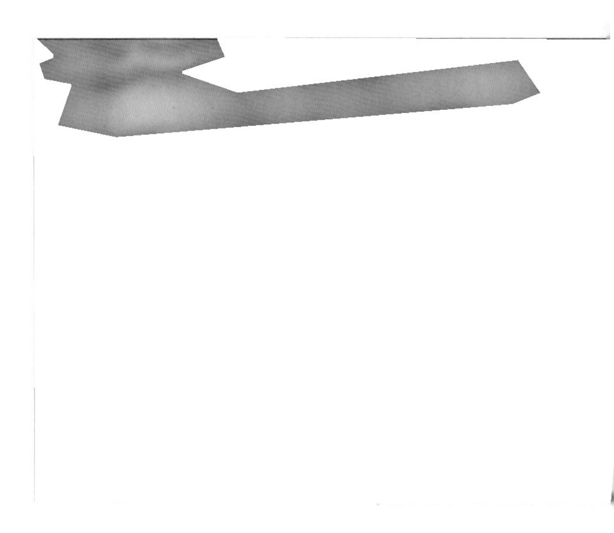

Ingredients

Fruit	Quantity (Crushed)	Sugar	Lemon Juice	Yield (Half-pints)
Apricots	4 to 4½ cups	4 cups	2 tbsps	5 to 6
Berries*	4 cups	4 cups	None	3 to 4
Peaches	5½ to 6 cups	4 to 5 cups	2 tbsps	6 to 7

* Includes blackberries, boysenberries, dewberries, gooseberries, loganberries, raspberries, and strawberries.

Temperature test—Use a jelly or candy thermometer and boil until mixture reaches the temperature for your altitude.

Sea Level	1,000 ft	2,000 ft	3,000 ft	4,000 ft	5,000 ft	6,000 ft	7,000 ft	8,000 ft
220°F	218°F	216°F	214°F	212°F	211°F	209°F	207°F	205°F

Refrigerator test—Remove the jam mixture from the heat. Pour a small amount of boiling jam on a cold plate and put it in the freezer compartment of a refrigerator for a few minutes. If the mixture gels, it is ready to fill.

PROCESS TIMES FOR JAMS WITHOUT ADDED PECTIN IN A BOILING-WATER CANNER*

		Process Time at Altitudes of:		
Style of Pack	Jar Size	0–1,000 ft	1,001–6,000 ft	Above 6,000 ft
Hot	Half-pints	5 minutes	10 minutes	15 minutes
*After the process is complete, turn off the heat and remove the canner lid. Wait five minutes before removing jars.				

2. Add sugar and bring to a boil while stirring rapidly and constantly. Continue to boil until mixture thickens. Use one of the tests on the previous page to determine when jams and jellies are ready to fill. Remember that the jam will thicken as it cools.
3. Remove from heat and skim off foam quickly. Fill sterile jars with jam. Use a measuring cup or ladle the jam through a wide-mouthed funnel, leaving ¼-inch headspace. Adjust lids and process.

Jams and Jellies with Added Pectin

To use fresh fruits with a low pectin content or canned or frozen fruit juice, powdered or liquid pectin must be added for your jams and jellies to thicken and set properly. Jelly or jam made with added pectin requires less cooking and generally gives a larger yield. These products have more natural fruit flavors, too. In addition, using added pectin eliminates the need to test hot jellies and jams for proper gelling.

Commercially produced pectin is a natural ingredient, usually made from apples and available at most grocery stores. There are several types of pectin now commonly available; liquid, powder, low-sugar, and no-sugar pectins each have their own advantages and downsides. Pomona's Universal Pectin® is a citrus pectin that allows you to make jams and jellies with little or no sugar. Because the order of combining ingredients depends on the type of pectin used, it is best to follow the common jam and jelly recipes that are included right on most pectin packages. However, if you want to try something a little different, follow one of the following recipes for mixed fruit and spiced fruit jams and jellies.

Pear-Apple Jam

This is a delicious jam perfect for making at the end of autumn, just before the frost gets the last apples. For a warming, spicy twist, add a teaspoon of fresh grated ginger along with the cinnamon. This recipe yields 7 to 8 half-pints.

Ingredients

- 2 cups peeled, cored, and finely chopped pears (about 2 lbs.)
- 1 cup peeled, cored, and finely chopped apples
- ¼ tsp ground cinnamon
- 6½ cups sugar
- 1/3 cup bottled lemon juice
- 6 oz liquid pectin

PROCESS TIMES FOR JAMS AND JELLIES WITH ADDED PECTIN IN A BOILING-WATER CANNER*

		Process Time at Altitudes of:		
Style of Pack	Jar Size	0–1,000 ft	1,001–6,000 ft	Above 6,000 ft
Hot	Half-pints	5 minutes	10 minutes	15 minutes

*After the process is complete, turn off the heat and remove the canner lid. Wait five minutes before removing jars.

TIP

- Adding ½ teaspoon of butter or margarine with the juice and pectin will reduce foaming. However, these may cause off-flavor in a long-term storage of jellies and jams.
- Purchase fresh fruit pectin each year. Old pectin may result in poor gels.
- Be sure to use mason canning jars, self-sealing two-piece lids, and a five-minute process (corrected for altitude, as necessary) in boiling water.

PROCESS TIMES FOR PEAR-APPLE JAM IN A BOILING WATER CANNER*

		Process Time at Altitudes of:		
Style of Pack	Jar Size	0–1,000 ft	1,001–6,000 ft	Above 6,000 ft
Hot	Half-pints	5 minutes	10 minutes	15 minutes

*After the process is complete, turn off the heat and remove the canner lid. Wait five minutes before removing jars.

Directions

1. Peel, core, and slice apples and pears into a large saucepan and stir in cinnamon. Thoroughly mix sugar and lemon juice with fruits and bring to a boil over high heat, stirring constantly and crushing fruit with a potato masher as it softens.
2. Once boiling, immediately stir in pectin. Bring to a full rolling boil and boil hard 1 minute, stirring constantly.
3. Remove from heat, quickly skim off foam, and fill sterile jars, leaving ¼-inch headspace. Adjust lids and process.

Strawberry-Rhubarb Jelly

Strawberry-rhubarb jelly will turn any ordinary piece of bread into a delightful treat. You can also spread it on shortcake or pound cake for a simple and unique dessert. This recipe yields about 7 half-pints.

Ingredients

- 1½ lbs red stalks of rhubarb
- 1½ qts ripe strawberries
- ½ tsp butter or margarine to reduce foaming (optional)
- 6 cups sugar
- 6 oz liquid pectin

Directions

1. Wash and cut rhubarb into 1-inch pieces and blend or grind. Wash, stem, and crush strawberries, one layer at a time, in a saucepan. Place both fruits in a jelly bag or double layer of cheesecloth and gently squeeze juice into a large measuring cup or bowl.
2. Measure 3½ cups of juice into a large saucepan. Add butter and sugar, thoroughly mixing into juice. Bring to a boil over high heat, stirring constantly.

3. As soon as mixture begins to boil, stir in pectin. Bring to a full rolling boil and boil hard 1 minute, stirring constantly. Remove from heat, quickly skim off foam, and fill sterile jars, leaving ¼-inch headspace. Adjust lids and process.

Blueberry-Spice Jam

This is a summery treat that is delicious spread over waffles with a little butter. Using wild blueberries results in a stronger flavor, but cultivated blueberries also work well. This recipe yields about five half-pints.

Ingredients

- 2½ pints ripe blueberries
- 1 tbsp lemon juice
- ½ tsp ground nutmeg or cinnamon
- ¾ cup water
- 5½ cups sugar
- 1 box (1¾ oz) powdered pectin

Directions

1. Wash and thoroughly crush blueberries, adding one layer at a time, in a saucepan. Add lemon juice, spice, and water. Stir pectin and bring to a full, rolling boil over high heat, stirring frequently.
2. Add the sugar and return to a full rolling boil. Boil hard for 1 minute, stirring constantly. Remove from heat, quickly skim off foam, and fill sterile jars, leaving ¼-inch headspace. Adjust lids and process.

Grape-Plum Jelly

If you think peanut butter and jelly sandwiches are only for kids, try grape-plum jelly spread with a natural nut butter over a thick slice of whole wheat bread. You'll change your mind. This recipe yields about 10 half-pints.

Ingredients

- 3½ lbs ripe plums
- 3 lbs ripe Concord grapes

PROCESS TIMES FOR STRAWBERRY-RHUBARB JELLY IN A BOILING-WATER CANNER*

		Process Time at Altitudes of:		
Style of Pack	Jar Size	0–1,000 ft	1,001–6,000 ft	Above 6,000 ft
Hot	Half-pints or pints	5 minutes	10 minutes	15 minutes

*After the process is complete, turn off the heat and remove the canner lid. Wait five minutes before removing jars.

PROCESS TIMES FOR BLUEBERRY-SPICE JAM IN A BOILING-WATER CANNER*

		Process Time at Altitudes of:		
Style of Pack	Jar Size	0–1,000 ft	1,001–6,000 ft	Above 6,000 ft
Hot	Half-pints or pints	5 minutes	10 minutes	15 minutes

*After the process is complete, turn off the heat and remove the canner lid. Wait five minutes before removing jars.

- 8½ cups sugar
- 1 cup water
- ½ tsp butter or margarine to reduce foaming (optional)
- 1 box (1¾ oz) powdered pectin

Directions

1. Wash and pit plums; do not peel. Thoroughly crush the plums and grapes, adding one layer at a time, in a saucepan with water. Bring to a boil, cover, and simmer ten minutes.
2. Strain juice through a jelly bag or double layer of cheesecloth. Measure sugar and set aside. Combine 6½ cups of juice with butter and pectin in large saucepan. Bring to a hard boil over high heat, stirring constantly.
3. Add the sugar and return to a full rolling boil. Boil hard for 1 minute, stirring constantly. Remove from heat, quickly skim off foam, and fill sterile jars, leaving ¼-inch headspace. Adjust lids and process.

Making Reduced-sugar Fruit Spreads

A variety of fruit spreads may be made that are tasteful, yet lower in sugars and calories than regular jams and jellies. The most straightforward method is probably to buy low-sugar pectin and follow the directions

on the package, but the recipes below show alternate methods of using gelatin or fruit pulp as thickening agents. Gelatin recipes should not be processed and should be refrigerated and used within four weeks.

Peach-Pineapple Spread

This recipe may be made with any combination of peaches, nectarines, apricots, and plums. You can use no sugar, up to two cups of sugar, or a combination of sugar and another sweetener (such as honey, Splenda®, or agave nectar). Note that if you use aspartame, the spread may lose its sweetness within three to four weeks. Add cinnamon or star anise if desired. This recipe yields 5 to 6 half-pints.

Ingredients

- 4 cups drained peach pulp (follow directions below)
- 2 cups drained, unsweetened crushed pineapple
- ¼ cup bottled lemon juice
- 2 cups sugar (optional)

PROCESS TIMES FOR GRAPE-PLUM JELLY IN A BOILING-WATER CANNER*

		Process Time at Altitudes of:		
Style of Pack	Jar Size	0–1,000 ft	1,001–6,000 ft	Above 6,000 ft
Hot	Half-pints or Pints	5 minutes	10 minutes	15 minutes

*After the process is complete, turn off the heat and remove the canner lid. Wait five minutes before removing jars.

PROCESS TIMES FOR PEACH-PINEAPPLE SPREAD IN A BOILING-WATER CANNER*

		Process Time at Altitudes of:			
Style of Pack	Jar Size	0–1,000 ft	1,001–3,000 ft	3,001–6,000 ft	Above 6,000 ft
Hot	Half-pints	15 minutes	20 minutes	20 minutes	25 minutes
	Pints	20 minutes	25 minutes	30 minutes	35 minutes

*After the process is complete, turn off the heat and remove the canner lid. Wait five minutes before removing jars.

Directions

1. Thoroughly wash 4 to 6 pounds of firm, ripe peaches. Drain well. Peel and remove pits. Grind fruit flesh with a medium or coarse blade, or crush with a fork (do not use a blender).
2. Place ground or crushed peach pulp in a 2-quart saucepan. Heat slowly to release juice, stirring constantly, until fruit is tender. Place cooked fruit in a jelly bag or strainer lined with four layers of cheesecloth. Allow juice to drip about 15 minutes. Save the juice for jelly or other uses.
3. Measure 4 cups of drained peach pulp for making spread. Combine the 4 cups of pulp, pineapple, and lemon juice in a 4-quart saucepan. Add up to 2 cups of sugar or other sweetener, if desired, and mix well.
4. Heat and boil gently for 10 to 15 minutes, stirring enough to prevent sticking. Fill jars quickly, leaving ¼-inch headspace. Adjust lids and process.

Refrigerated Apple Spread

This recipe uses gelatin as a thickener, so it does not require processing but it should be refrigerated and used within four weeks. For spiced apple jelly, add two sticks of cinnamon and four whole cloves to mixture before boiling. Remove both spices before adding the sweetener and food coloring (if desired). This recipe yields 4 half-pints.

Ingredients

- 2 tbsps unflavored gelatin powder
- 1 qt bottle unsweetened apple juice
- 2 tbsps bottled lemon juice
- 2 tbsps liquid low-calorie sweetener (e.g., sucralose, honey, or 1–2 tsps liquid stevia)

Directions

1. In a saucepan, soften the gelatin in the apple and lemon juices. To dissolve gelatin, bring to a full rolling boil and boil 2 minutes. Remove from heat.
2. Stir in sweetener and food coloring (if desired). Fill jars, leaving ¼-inch headspace. Adjust lids. Refrigerate (do not process or freeze).

Refrigerated Grape Spread

This is a simple, tasty recipe that doesn't require processing. Be sure to refrigerate and use within four weeks. This recipe makes 3 half-pints.

Ingredients

- 2 tbsps unflavored gelatin powder
- 1 bottle (24 oz) unsweetened grape juice
- 2 tbsps bottled lemon juice
- 2 tbsps liquid low-calorie sweetener (e.g., sucralose, honey, or 1–2 tsps liquid stevia)

Directions

1. In a saucepan, heat the gelatin in the grape and lemon juices until mixture is soft. Bring to a full rolling boil to dissolve gelatin. Boil 1 minute and remove from heat. Stir in sweetener.
2. Fill jars quickly, leaving ¼-inch headspace. Adjust lids. Refrigerate (do not process or freeze).

Remaking Soft Jellies

Sometimes jelly just doesn't turn out right the first time. Jelly that is too soft can be used as a sweet sauce to drizzle over ice cream, cheesecake, or angel food cake, but it can also be recooked into the proper consistency.

To Remake with Powdered Pectin

1. Measure jelly to be recooked. Work with no more than 4 to 6 cups at a time. For each quart (4 cups) of jelly, mix ¼ cup sugar, ½ cup water, 2 tablespoons bottled lemon juice, and 4 teaspoons powdered pectin. Bring to a boil while stirring.
2. Add jelly and bring to a rolling boil over high heat, stirring constantly. Boil hard ½ minute. Remove from heat, quickly skim foam off jelly, and fill sterile jars, leaving ¼-inch headspace. Adjust new lids and process as recommended (see next page).

To Remake with Liquid Pectin

1. Measure jelly to be recooked. Work with no more than 4 to 6 cups at a time. For each quart (4 cups) of jelly, measure into a bowl ¾ cup sugar, 2 tablespoons bottled lemon juice, and 2 tablespoons liquid pectin.
2. Bring jelly only to boil over high heat, while stirring. Remove from heat and quickly add the sugar, lemon juice, and pectin. Bring to a full rolling boil, stirring constantly. Boil hard for 1 minute. Quickly skim off foam and fill sterile jars, leaving ¼-inch headspace. Adjust new lids and process as recommended (see page next page).

To Remake without Added Pectin

1. For each quart of jelly, add 2 tablespoons bottled lemon juice. Heat to boiling and continue to boil for 3 to 4 minutes.
2. To test jelly for doneness, use one of the following methods on the next page.
3. Remove from heat, quickly skim off foam, and fill sterile jars, leaving ¼-inch headspace. Adjust new lids and process.

Vegetables, Pickles, and Tomatoes

Beans or Peas, Shelled or Dried (All Varieties)

Shelled or dried beans and peas are inexpensive and easy to buy or store in bulk, but they are not very convenient when it comes to preparing them to eat. Hydrating and canning beans or peas enable you to simply open a can and use them rather than waiting for them to soak. Sort and discard discolored seeds before rehydrating.

Temperature test—Use a jelly or candy thermometer and boil until mixture reaches the following temperatures at the altitudes below:

Sea Level	1,000 ft	2,000 ft	3,000 ft	4,000 ft	5,000 ft	6,000 ft	7,000 ft	8,000 ft
220°F	218°F	216°F	214°F	212°F	211°F	209°F	207°F	205°F

Sheet or spoon test—Dip a cool metal spoon into the boiling jelly mixture. Raise the spoon about 12 inches above the pan (out of steam). Turn the spoon so the liquid runs off the side. The jelly is done when the syrup forms two drops that flow together and sheet or hang off the edge of the spoon.

PROCESS TIMES FOR REMADE SOFT JELLIES IN A BOILING-WATER CANNER*

		Process Time at Altitudes of:		
Style of Pack	Jar Size	0–1,000 ft	1,001–6,000 ft	Above 6,000 ft
Hot	Half-pints or Pints	5 minutes	10 minutes	15 minutes

*After the process is complete, turn off the heat and remove the canner lid. Wait five minutes before removing jars.

Directions

1. Place dried beans or peas in a large pot and cover with water. Soak 12 to 18 hours in a cool place. Drain water. To quickly hydrate beans, you may cover sorted and washed beans with boiling water in a saucepan. Boil 2 minutes, remove from heat, soak 1 hour, and drain.
2. Cover beans soaked by either method with fresh water and boil 30 minutes. Add ½ teaspoon of salt per pint or 1 teaspoon per quart to each jar, if desired. Fill jars with beans or peas and cooking water, leaving 1-inch headspace. Adjust lids and process.

QUANTITY

- An average of 5 pounds of beans is needed per canner load of 7 quarts.
- An average of 3¼ pounds is needed per canner load of 9 pints—an average of ¾ pounds per quart.

Baked Beans

Baked beans are an old New England favorite, but every cook has his or her favorite variation. Two recipes are included here, but feel free to alter them to your own taste.

Directions

1. Sort and wash dry beans. Add 3 cups of water for each cup of dried beans. Boil 2 minutes, remove from heat, soak 1 hour, and drain.
2. Heat to boiling in fresh water, and save liquid for making sauce. Make your choice of the following sauces:

PROCESS TIMES FOR BEANS OR PEAS IN A DIAL-GAUGE PRESSURE CANNER*

			Canner Pressure (PSI) at Altitudes of:			
Style of Pack	Jar Size	Process Time	0–2,000 ft	2,001–4,000 ft	4,001–6,000 ft	6,001–8,000 ft
Hot	Pints	75 minutes	11 lbs	12 lbs	13 lbs	14 lbs
	Quarts	90 minutes	11 lbs	12 lbs	13 lbs	14 lbs
*After the canner is completely depressurized, remove the weight from the vent port or open the petcock. Wait ten minutes; then unfasten the lid and remove it carefully. Lift the lid with the underside away from you so that the steam coming out of the canner does not burn your face.						

PROCESS TIMES FOR BEANS OR PEAS IN A WEIGHTED-GAUGE PRESSURE CANNER*

			Canner Pressure (PSI) at Altitudes of:	
Style of pack	Jar Size	Process Time	0–1,000 ft	Above 1,000 ft
Hot	Pints	75 minutes	10 lbs	15 lbs
	Quarts	90 minutes	10 lbs	15 lbs
*After the canner is completely depressurized, remove the weight from the vent port or open the petcock. Wait ten minutes; then unfasten the lid and remove it carefully. Lift the lid with the underside away from you so that the steam coming out of the canner does not burn your face.				

Tomato Sauce—Mix 1 quart tomato juice, 3 tablespoons sugar, 2 teaspoons salt, 1 tablespoon chopped onion, and ¼ teaspoon each of ground cloves, allspice, mace, and cayenne pepper. Heat to boiling. Add 3 quarts cooking liquid from beans and bring back to boiling.

Molasses Sauce—Mix 4 cups water or cooking liquid from beans, 3 tablespoons dark molasses, 1 tablespoon vinegar, 2 teaspoons salt, and ¾ teaspoon powdered dry mustard. Heat to boiling.

QUANTITY

- An average of 5 pounds of beans is needed per canner load of 7 quarts.
- An average of 3¼ pounds is needed per canner load of 9 pints—an average of ¾ pounds per quart.

3. Place seven ¾-inch pieces of pork, ham, or bacon in an earthenware crock, a large casserole, or a pan. Add beans and enough molasses sauce to cover beans.
4. Cover and bake 4 to 5 hours at 350°F. Add water as needed—about every hour. Fill jars, leaving 1-inch headspace. Adjust lids and process.

Green Beans

This process will work equally well for snap, Italian, or wax beans. Select filled but tender, crisp pods, removing any diseased or rusty pods.

Directions

1. Wash beans and trim ends. Leave whole, or cut, or break into 1-inch pieces.

Hot pack—Cover with boiling water; boil 5 minutes. Fill jars loosely, leaving 1-inch headspace.

PROCESS TIMES FOR BAKED BEANS IN A DIAL-GAUGE PRESSURE CANNER*

			Canner Pressure (PSI) at Altitudes of:			
Style of Pack	Jar Size	Process Time	0–2,000 ft	2,001–4,000 ft	4,001–6,000 ft	6,001–8,000 ft
Hot	Pints	65 minutes	11 lbs	12 lbs	13 lbs	14 lbs
	Quarts	75 minutes	11 lbs	12 lbs	13 lbs	14 lbs

*After the canner is completely depressurized, remove the weight from the vent port or open the petcock. Wait ten minutes; then unfasten the lid and remove it carefully. Lift the lid with the underside away from you so that the steam coming out of the canner does not burn your face.

PROCESS TIMES FOR BAKED BEANS IN A WEIGHTED-GAUGE PRESSURE CANNER*

			Canner Pressure (PSI) at Altitudes of:	
Style of pack	Jar Size	Process Time	0–1,000 ft	Above 1,000 ft
Hot	Pints	65 minutes	10 lbs	15 lbs
	Quarts	75 minutes	10 lbs	15 lbs

*After the canner is completely depressurized, remove the weight from the vent port or open the petcock. Wait ten minutes; then unfasten the lid and remove it carefully. Lift the lid with the underside away from you so that the steam coming out of the canner does not burn your face.

Raw pack—Fill jars tightly with raw beans, leaving 1-inch headspace. Add 1 teaspoon of salt per quart to each jar, if desired. Add boiling water, leaving 1-inch headspace.

2. Adjust lids and process.

Beets

You can preserve beets whole, cubed, or sliced, according to your preference. Beets that are 1 to 2 inches in diameter are the best, as larger ones tend to be too fibrous.

Directions

1. Trim off beet tops, leaving an inch of stem and roots to reduce bleeding of color. Scrub well. Cover with boiling water. Boil until skins slip off easily, about 15 to 25 minutes depending on size.
2. Cool, remove skins, and trim off stems and roots. Leave baby beets whole. Cut medium or large beets into ½-inch cubes

QUANTITY

- An average of 14 pounds of beans is needed per canner load of 7 quarts.
- An average of nine pounds is needed per canner load of 9 pints.
- A bushel weighs 30 pounds and yields 12 to 20 quarts—an average of 2 pounds per quart.

PROCESS TIMES FOR GREEN BEANS IN A DIAL-GAUGE PRESSURE CANNER*

			Canner Pressure (PSI) at Altitudes of:			
Style of Pack	Jar Size	Process Time	0–2,000 ft	2,001–4,000 ft	4,001–6,000 ft	6,001–8,000 ft
Hot or Raw	Pints	20 minutes	11 lbs	12 lbs	13 lbs	14 lbs
	Quarts	25 minutes	11 lbs	12 lbs	13 lbs	14 lbs

*After the canner is completely depressurized, remove the weight from the vent port or open the petcock. Wait ten minutes; then unfasten the lid and remove it carefully. Lift the lid with the underside away from you so that the steam coming out of the canner does not burn your face.

PROCESS TIMES FOR GREEN BEANS IN A WEIGHTED-GAUGE PRESSURE CANNER*

			Canner Pressure (PSI) at Altitudes of:	
Style of Pack	Jar Size	Process Time	0–1,000 ft	Above 1,000 ft
Hot or Raw	Pints	20 minutes	10 lbs	15 lbs
	Quarts	25 minutes	10 lbs	15 lbs

*After the canner is completely depressurized, remove the weight from the vent port or open the petcock. Wait ten minutes; then unfasten the lid and remove it carefully. Lift the lid with the underside away from you so that the steam coming out of the canner does not burn your face.

or slices. Halve or quarter very large slices. Add 1 teaspoon of salt per quart to each jar, if desired.

3. Fill jars with hot beets and fresh hot water, leaving 1-inch headspace. Adjust lids and process.

Carrots

Carrots can be preserved sliced or diced according to your preference. Choose small carrots, preferably 1 to 1¼ inches in diameter, as larger ones are often too fibrous.

Directions

1. Wash, peel, and rewash carrots. Slice or dice.

Hot pack—Cover with boiling water; bring to boil and simmer for 5 minutes. Fill jars with carrots, leaving 1-inch headspace.

Raw pack—Fill jars tightly with raw carrots, leaving 1-inch headspace.

QUANTITY

- An average of 21 pounds of beets (without tops) is needed per canner load of 7 quarts.
- An average of 13½ pounds is needed per canner load of 9 pints.
- A bushel (without tops) weighs 52 pounds and yields 15 to 20 quarts—an average of 3 pounds per quart.

QUANTITY

- An average of 17½ pounds of carrots (without tops) is needed per canner load of 7 quarts.
- An average of 11 pounds is needed per canner load of 9 pints.
- A bushel (without tops) weighs 50 pounds and yields 17 to 25 quarts—an average of 2½ pounds per quart.

2. Add 1 teaspoon of salt per quart to the jar, if desired. Add hot cooking liquid or water, leaving 1-inch headspace. Adjust lids and process.

Corn, Cream Style

The creamy texture comes from scraping the corncobs thoroughly and including the juices and corn pieces with the kernels. If you want to add milk or cream, butter, or other ingredients, do so just before serving (do not add dairy products before canning). Select ears containing slightly immature kernels for this recipe.

QUANTITY

- An average of 20 pounds (in husks) of sweet corn is needed per canner load of 9 pints.
- A bushel weighs 35 pounds and yields 12 to 20 pints—an average of 2¼ pounds per pint.

Directions

1. Husk corn, remove silk, and wash ears. Cut corn from cob at about the center of kernel. Scrape remaining corn from cobs with a table knife.

Hot pack—To each quart of corn and scrapings in a saucepan, add 2 cups of boiling water. Heat to boiling. Add ½ teaspoon salt to each jar, if desired. Fill pint jars with hot corn mixture, leaving 1-inch headspace.

Raw pack—Fill pint jars with raw corn, leaving 1-inch headspace. Do not shake or press down. Add ½ teaspoon salt to each jar,

PROCESS TIMES FOR BEETS IN A DIAL-GAUGE PRESSURE CANNER*

			Canner Pressure (PSI) at Altitudes of:			
Style of Pack	Jar Size	Process Time	0–2,000 ft	2,001–4,000 ft	4,001–6,000 ft	6,001–8,000 ft
Hot	Pints	30 minutes	11 lbs	12 lbs	13 lbs	14 lbs
	Quarts	35 minutes	11 lbs	12 lbs	13 lbs	14 lbs

*After the canner is completely depressurized, remove the weight from the vent port or open the petcock. Wait ten minutes; then unfasten the lid and remove it carefully. Lift the lid with the underside away from you so that the steam coming out of the canner does not burn your face.

PROCESS TIMES FOR BEETS IN A WEIGHTED-GAUGE PRESSURE CANNER*

			Canner Pressure (PSI) at Altitudes of:	
Style of Pack	Jar Size	Process Time	0–1,000 ft	Above 1,000 ft
Hot or Raw	Pints	30 minutes	10 lbs	15 lbs
	Quarts	35 minutes	10 lbs	15 lbs

*After the canner is completely depressurized, remove the weight from the vent port or open the petcock. Wait ten minutes; then unfasten the lid and remove it carefully. Lift the lid with the underside away from you so that the steam coming out of the canner does not burn your face.

PROCESS TIMES FOR CARROTS IN A DIAL-GAUGE PRESSURE CANNER*

			Canner Pressure (PSI) at Altitudes of:			
Style of Pack	Jar Size	Process Time	0–2,000 ft	2,001–4,000 ft	4,001–6,000 ft	6,001–8,000 ft
Hot or Raw	Pints	25 minutes	11 lbs	12 lbs	13 lbs	14 lbs
	Quarts	30 minutes	11 lbs	12 lbs	13 lbs	14 lbs

*After the canner is completely depressurized, remove the weight from the vent port or open the petcock. Wait ten minutes; then unfasten the lid and remove it carefully. Lift the lid with the underside away from you so that the steam coming out of the canner does not burn your face.

PROCESS TIMES FOR CARROTS IN A WEIGHTED-GAUGE PRESSURE CANNER*

			Canner Pressure (PSI) at Altitudes of:	
Style of Pack	Jar Size	Process Time	0–1,000 ft	Above 1,000 ft
Hot or Raw	Pints	25 minutes	10 lbs	15 lbs
	Quarts	30 minutes	10 lbs	15 lbs

*After the canner is completely depressurized, remove the weight from the vent port or open the petcock. Wait ten minutes; then unfasten the lid and remove it carefully. Lift the lid with the underside away from you so that the steam coming out of the canner does not burn your face.

if desired. Add fresh boiling water, leaving 1-inch headspace.

2. Adjust lids and process.

Corn, Whole Kernel

Select ears containing slightly immature kernels. Canning of some sweeter varieties or kernels that are too immature may cause browning. Try canning a small amount to test color and flavor before canning large quantities.

QUANTITY

- An average of 31½ pounds (in husks) of sweet corn is needed per canner load of 7 quarts.
- An average of 20 pounds is needed per canner load of 9 pints.
- A bushel weighs 35 pounds and yields 6 to 11 quarts—an average of 4½ pounds per quart.

Directions

1. Husk corn, remove silk, and wash. Blanch 3 minutes in boiling water. Cut corn from cob at about three-fourths the depth of kernel. Do not scrape cob, as it will create a creamy texture.

Hot pack—To each quart of kernels in a saucepan, add 1 cup of hot water, heat to boiling, and simmer 5 minutes. Add 1 teaspoon of salt per quart to each jar, if desired. Fill jars with corn and cooking liquid, leaving 1-inch headspace.

Raw pack—Fill jars with raw kernels, leaving 1-inch headspace. Do not shake or press down. Add 1 teaspoon of salt per quart to the jar, if desired.

2. Add fresh boiling water, leaving 1-inch headspace. Adjust lids and process.

Mixed Vegetables

Use mixed vegetables in soups, casseroles, pot pies, or as a quick side dish. You can change the suggested proportions or substitute other favorite vegetables, but avoid leafy greens, dried beans, cream-style corn, winter

PROCESS TIMES FOR CREAM-STYLE CORN IN A DIAL-GAUGE PRESSURE CANNER

			Canner Pressure (PSI) at Altitudes of:			
Style of pack	Jar Size	Process Time	0–2,000 ft	2,001–4,000 ft	4,001–6,000 ft	6,001–8,000 ft
Hot	Pints	85 minutes	11 lbs	12 lbs	13 lbs	14 lbs
Raw	Pints	95 minutes	11 lbs	12 lbs	13 lbs	14 lbs

*After the canner is completely depressurized, remove the weight from the vent port or open the petcock. Wait ten minutes; then unfasten the lid and remove it carefully. Lift the lid with the underside away from you so that the steam coming out of the canner does not burn your face.

PROCESS TIMES FOR CREAM-STYLE CORN IN A WEIGHTED-GAUGE PRESSURE CANNER*

			Canner Pressure (PSI) at Altitudes of:	
Style of Pack	Jar Size	Process Time	0–1,000 ft	Above 1,000 ft
Hot	Pints	85 minutes	10 lbs	15 lbs
Raw	Pints	95 minutes	10 lbs	15 lbs

*After the canner is completely depressurized, remove the weight from the vent port or open the petcock. Wait ten minutes; then unfasten the lid and remove it carefully. Lift the lid with the underside away from you so that the steam coming out of the canner does not burn your face.

squash, and sweet potatoes as they will ruin the consistency of the other vegetables. This recipe yields about 7 quarts.

Ingredients

- 6 cups sliced carrots
- 6 cups cut, whole-kernel sweet corn
- 6 cups cut green beans
- 6 cups shelled lima beans
- 4 cups diced or crushed tomatoes
- 4 cups diced zucchini

Directions

1. Carefully wash, peel, de-shell, and cut vegetables as necessary. Combine all vegetables in a large pot or kettle, and add enough water to cover pieces.
2. Add 1 teaspoon salt per quart to each jar, if desired. Boil 5 minutes and fill jars with hot pieces and liquid, leaving 1-inch headspace. Adjust lids and process.

PROCESS TIMES FOR WHOLE KERNEL CORN IN A DIAL-GAUGE PRESSURE CANNER*

			Canner Pressure (PSI) at Altitudes of:			
Style of Pack	Jar Size	Process Time	0–2,000 ft	2,001–4,000 ft	4,001–6,000 ft	6,001–8,000 ft
Hot or Raw	Pints	55 minutes	11 lbs	12 lbs	13 lbs	14 lbs
	Quarts	85 minutes	11 lbs	12 lbs	13 lbs	14 lbs

*After the canner is completely depressurized, remove the weight from the vent port or open the petcock. Wait ten minutes; then unfasten the lid and remove it carefully. Lift the lid with the underside away from you so that the steam coming out of the canner does not burn your face.

PROCESS TIMES FOR MIXED VEGETABLES IN A DIAL-GAUGE PRESSURE CANNER*

			Canner Pressure (PSI) at Altitudes of:			
Style of Pack	Jar Size	Process Time	0–2,000 ft	2,001–4,000 ft	4,001–6,000 ft	6,001–8,000 ft
Hot	Pints	75 minutes	11 lbs	12 lbs	13 lbs	14 lbs
	Quarts	90 minutes	11 lbs	12 lbs	13 lbs	14 lbs

*After the canner is completely depressurized, remove the weight from the vent port or open the petcock. Wait ten minutes; then unfasten the lid and remove it carefully. Lift the lid with the underside away from you so that the steam coming out of the canner does not burn your face.

PROCESS TIMES FOR MIXED VEGETABLES IN A WEIGHTED-GAUGE PRESSURE CANNER*

			Canner Pressure (PSI) at Altitudes of:	
Style of Pack	Jar Size	Process Time	0–1,000 ft	Above 1,000 ft
Hot	Pints	75 minutes	10 lbs	15 lbs
	Quarts	90 minutes	10 lbs	15 lbs

*After the canner is completely depressurized, remove the weight from the vent port or open the petcock. Wait ten minutes; then unfasten the lid and remove it carefully. Lift the lid with the underside away from you so that the steam coming out of the canner does not burn your face.

Peas, Green or English, Shelled

Green and English peas preserve well when canned, but sugar snap and Chinese edible pods are better frozen. Select filled pods containing young, tender, sweet seeds, and discard any diseased pods.

Directions

1. Shell and wash peas. Add 1 teaspoon of salt per quart to each jar, if desired.

Hot pack—Cover with boiling water. Bring to a boil in a saucepan, and boil 2 minutes. Fill jars loosely with hot peas, and add cooking liquid, leaving 1-inch headspace.

Raw pack—Fill jars with raw peas, and add boiling water, leaving 1-inch headspace. Do not shake or press down peas.

2. Adjust lids and process.

QUANTITY

- An average of 31½ pounds of peas (in pods) is needed per canner load of 7 quarts.
- An average of 20 pounds is needed per canner load of 9 pints.
- A bushel weighs 30 pounds and yields 5 to 10 quarts—an average of 4½ pounds per quart.

Potatoes, Sweet

Sweet potatoes can be preserved whole, in chunks, or in slices, according to your preference. Choose small to medium-sized potatoes that are mature and not too fibrous. Can within one to two months after harvest.

QUANTITY

- An average of 17½ pounds of sweet potatoes is needed per canner load of 7 quarts.
- An average of 11 pounds is needed per canner load of 9 pints.
- A bushel weighs 50 pounds and yields 17 to 25 quarts—an average of 2½ pounds per quart.

PROCESS TIME FOR PEAS IN A DIAL-GAUGE PRESSURE CANNER*

			Canner Pressure (PSI) at Altitudes of:			
Style of Pack	Jar Size	Process Time	0–2,000 ft	2,001–4,000 ft	4,001–6,000 ft	6,001–8,000 ft
Hot or Raw	Pints or Quarts	40 minutes	11 lbs	12 lbs	13 lbs	14 lbs

*After the canner is completely depressurized, remove the weight from the vent port or open the petcock. Wait ten minutes; then unfasten the lid and remove it carefully. Lift the lid with the underside away from you so that the steam coming out of the canner does not burn your face.

PROCESS TIME FOR PEAS IN A WEIGHTED-GAUGE PRESSURE CANNER*

			Canner Pressure (PSI) at Altitudes of:	
Style of Pack	Jar Size	Process Time	0–1,000 ft	Above 1,000 ft
Hot or Raw	Pints or Quarts	40 minutes	10 lbs	15 lbs

*After the canner is completely depressurized, remove the weight from the vent port or open the petcock. Wait ten minutes; then unfasten the lid and remove it carefully. Lift the lid with the underside away from you so that the steam coming out of the canner does not burn your face.

PROCESS TIMES FOR SWEET POTATOES IN A DIAL-GAUGE PRESSURE CANNER*

			Canner Pressure (PSI) at Altitudes of:			
Style of Pack	Jar Size	Process Time	0–2,000 ft	2,001–4,000 ft	4,001–6,000 ft	6,001–8,000 ft
Hot	Pints	65 minutes	11 lbs	12 lbs	13 lbs	14 lbs
	Quarts	90 minutes	11 lbs	12 lbs	13 lbs	14 lbs

*After the canner is completely depressurized, remove the weight from the vent port or open the petcock. Wait ten minutes; then unfasten the lid and remove it carefully. Lift the lid with the underside away from you so that the steam coming out of the canner does not burn your face.

PROCESS TIMES FOR SWEET POTATOES IN A WEIGHTED-GAUGE PRESSURE CANNER*

			Canner Pressure (PSI) at Altitudes of:	
Style of Pack	Jar Size	Process Time	0–1,000 ft	Above 1,000 ft
Hot	Pints	65 minutes	10 lbs	15 lbs
	Quarts	90 minutes	10 lbs	15 lbs

*After the canner is completely depressurized, remove the weight from the vent port or open the petcock. Wait ten minutes; then unfasten the lid and remove it carefully. Lift the lid with the underside away from you so that the steam coming out of the canner does not burn your face.

Directions

1. Wash potatoes and boil or steam until partially soft (15 to 20 minutes). Remove skins. Cut medium potatoes, if needed, so that pieces are uniform in size. Do not mash or purée pieces.
2. Fill jars, leaving 1-inch headspace. Add 1 teaspoon salt per quart to each jar, if desired. Cover with your choice of fresh boiling water or syrup, leaving 1-inch headspace. Adjust lids and process.

Pumpkin and Winter Squash

Pumpkin and squash are great to have on hand for use in pies, soups, quick breads, or as side dishes. They should have a hard rind and stringless, mature pulp. Small pumpkins (sugar or pie varieties) are best. Before using for pies, drain jars and strain or sieve pumpkin or squash cubes.

Directions

1. Wash, remove seeds, cut into 1-inch-wide slices, and peel. Cut flesh into 1-inch cubes. Boil 2 minutes in water. Do not mash or purée.
2. Fill jars with cubes and cooking liquid, leaving 1-inch headspace. Adjust lids and process.

> **QUANTITY**
>
> - An average of 16 pounds of pumpkin is needed per canner load of 7 quarts.
> - An average of 10 pounds is needed per canner load of 9 pints—an average of 2¼ pounds per quart.

Succotash

To spice up this simple, satisfying dish, add a little paprika and celery salt before serving. It is also delicious made into a pot pie, with or without added chicken, turkey, or beef. This recipe yields seven quarts.

Ingredients

- 1 lb unhusked sweet corn or 3 qts cut whole kernels
- 14 lbs mature green podded lima beans or 4 qts shelled lima beans
- 2 qts crushed or whole tomatoes (optional)

Directions

1. Husk corn, remove silk, and wash. Blanch 3 minutes in boiling water. Cut corn from cob at about three-fourths the depth of kernel. Do not scrape cob, as it will create a creamy texture. Shell lima beans and wash thoroughly.

Hot pack—Combine all prepared vegetables in a large kettle with enough water to cover the pieces. Add 1 teaspoon salt to each quart jar, if desired. Boil gently 5 minutes and fill jars with pieces and cooking liquid, leaving 1-inch headspace.

Raw pack—Fill jars with equal parts of all prepared vegetables, leaving 1-inch headspace. Do not shake or press down pieces. Add 1 teaspoon salt to each quart jar, if desired. Add fresh boiling water, leaving 1-inch headspace.

2. Adjust lids and process.

Soups

Vegetable, dried bean or pea, meat, poultry, or seafood soups can all be canned. Add

(continued on page 136)

PROCESS TIMES FOR PUMPKIN AND WINTER SQUASH IN A DIAL-GAUGE PRESSURE CANNER*

			Canner Pressure (PSI) at Altitudes of:			
Style of Pack	Jar Size	Process Time	0–2,000 ft	2,001–4,000 ft	4,001–6,000 ft	6,001–8,000 ft
Hot	Pints	55 minutes	11 lbs	12 lbs	13 lbs	14 lbs
	Quarts	90 minutes	11 lbs	12 lbs	13 lbs	14 lbs

*After the canner is completely depressurized, remove the weight from the vent port or open the petcock. Wait ten minutes; then unfasten the lid and remove it carefully. Lift the lid with the underside away from you so that the steam coming out of the canner does not burn your face.

PROCESS TIMES FOR PUMPKIN AND WINTER SQUASH IN A WEIGHTED-GAUGE PRESSURE CANNER*

			Canner Pressure (PSI) at Altitudes of:	
Style of Pack	Jar Size	Process Time	0–1,000 ft	Above 1,000 ft
Hot	Pints	55 minutes	10 lbs	15 lbs
	Quarts	90 minutes	10 lbs	15 lbs

*After the canner is completely depressurized, remove the weight from the vent port or open the petcock. Wait ten minutes; then unfasten the lid and remove it carefully. Lift the lid with the underside away from you so that the steam coming out of the canner does not burn your face.

PROCESS TIMES FOR SUCCOTASH IN A DIAL-GAUGE PRESSURE CANNER*

			Canner Pressure (PSI) at Altitudes of:			
Style of Pack	Jar Size	Process Time	0–2,000 ft	2,001–4,000 ft	4,001–6,000 ft	6,001–8,000 ft
Hot or Raw	Pints	60 minutes	11 lbs	12 lbs	13 lbs	14 lbs
	Quarts	85 minutes	11 lbs	12 lbs	13 lbs	14 lbs

*After the canner is completely depressurized, remove the weight from the vent port or open the petcock. Wait ten minutes; then unfasten the lid and remove it carefully. Lift the lid with the underside away from you so that the steam coming out of the canner does not burn your face.

PROCESS TIMES FOR SUCCOTASH IN A WEIGHTED-GAUGE PRESSURE CANNER*

			Canner Pressure (PSI) at Altitudes of:	
Style of Pack	Jar Size	Process Time	0–1,000 ft	Above 1,000 ft
Hot or Raw	Pints	60 minutes	10 lbs	15 lbs
	Quarts	85 minutes	10 lbs	15 lbs

*After the canner is completely depressurized, remove the weight from the vent port or open the petcock. Wait ten minutes; then unfasten the lid and remove it carefully. Lift the lid with the underside away from you so that the steam coming out of the canner does not burn your face.

pasta, rice, or other grains to soup just prior to serving, as grains tend to get soggy when canned. If dried beans or peas are used, they *must* be fully rehydrated first. Dairy products should also be avoided in the canning process.

Directions

1. Select, wash, and prepare vegetables
2. Cook vegetables. For each cup of dried beans or peas, add 3 cups of water, boil 2 minutes, remove from heat, soak 1 hour, and heat to boil. Drain and combine with meat broth, tomatoes, or water to cover. Boil 5 minutes.
3. Salt to taste, if desired. Fill jars halfway with solid mixture. Add remaining liquid, leaving 1-inch headspace. Adjust lids and process.

Meat Stock (Broth)

"Good broth will resurrect the dead," says a South American proverb. Bones contain calcium, magnesium, phosphorus, and other trace minerals, while cartilage and tendons hold glucosamine, which is important for joints and muscle health. When simmered for extended periods, these nutrients are released into the water and broken down into a form that our bodies can absorb. Not to mention that good broth is the secret to delicious risotto, reduction sauces, gravies, and dozens of other gourmet dishes.

Beef

1. Saw or crack fresh trimmed beef bones to enhance extraction of flavor. Rinse bones and place in a large stockpot or kettle, cover bones with water, add pot cover, and simmer 3 to 4 hours.
2. Remove bones, cool broth, and pick off meat. Skim off fat, add meat removed from bones to broth, and reheat to boiling. Fill jars, leaving 1-inch headspace. Adjust lids and process.

Chicken or Turkey

1. Place large carcass bones in a large stockpot, add enough water to cover bones,

PROCESS TIMES FOR SOUPS IN A DIAL-GAUGE PRESSURE CANNER*

			Canner Pressure (PSI) at Altitudes of:			
Style of Pack	Jar Size	Process Time	0–2,000 ft	2,001–4,000 ft	4,001–6,000 ft	6,001–8,000 ft
Hot	Pints	60* minutes	11 lbs	12 lbs	13 lbs	14 lbs
	Quarts	75* minutes	11 lbs	12 lbs	13 lbs	14 lbs

***Caution: Process 100 minutes if soup contains seafood.**

*After the canner is completely depressurized, remove the weight from the vent port or open the petcock. Wait ten minutes; then unfasten the lid and remove it carefully. Lift the lid with the underside away from you so that the steam coming out of the canner does not burn your face.

PROCESS TIMES FOR SOUPS IN A WEIGHTED-GAUGE PRESSURE CANNER*

			Canner Pressure (PSI) at Altitudes of:	
Style of Pack	Jar Size	Process Time	0–1,000 ft	Above 1,000 ft
Hot	Pints	60* minutes	10 lbs	15 lbs
	Quarts	75* minutes	10 lbs	15 lbs

***Caution: Process 100 minutes if soup contains seafood.**

*After the canner is completely depressurized, remove the weight from the vent port or open the petcock. Wait ten minutes; then unfasten the lid and remove it carefully. Lift the lid with the underside away from you so that the steam coming out of the canner does not burn your face.

PROCESS TIMES FOR MEAT STOCK IN A DIAL-GAUGE PRESSURE CANNER*

			Canner Pressure (PSI) at Altitudes of:			
Style of Pack	Jar Size	Process Time	0–2,000 ft	2,001–4,000 ft	4,001–6,000 ft	6,001–8,000 ft
Hot	Pints	20 minutes	11 lbs	12 lbs	13 lbs	14 lbs
	Quarts	25 minutes	11 lbs	12 lbs	13 lbs	14 lbs

*After the canner is completely depressurized, remove the weight from the vent port or open the petcock. Wait ten minutes; then unfasten the lid and remove it carefully. Lift the lid with the underside away from you so that the steam coming out of the canner does not burn your face.

PROCESS TIMES FOR MEAT STOCK IN A WEIGHTED-GAUGE PRESSURE CANNER*

			Canner Pressure (PSI) at Altitudes of:	
Style of Pack	Jar Size	Process Time	0–1,000 ft	Above 1,000 ft
Hot	Pints	20 minutes	10 lbs	15 lbs
	Quarts	25 minutes	10 lbs	15 lbs

*After the canner is completely depressurized, remove the weight from the vent port or open the petcock. Wait ten minutes; then unfasten the lid and remove it carefully. Lift the lid with the underside away from you so that the steam coming out of the canner does not burn your face.

cover pot, and simmer 30 to 45 minutes or until meat can be easily stripped from bones.

2. Remove bones and pieces, cool broth, strip meat, discard excess fat, and return meat to broth. Reheat to boiling and fill jars, leaving 1-inch headspace. Adjust lids and process.

Fermented Foods and Pickled Vegetables

Pickled vegetables play a vital role in Italian antipasto dishes, Chinese stir-fries, British piccalilli, and much of Russian and Finnish cuisine. And, of course, the Germans love their sauerkraut, kimchee is found on nearly every Korean dinner table, and many an American won't eat a sandwich without a good strong dill pickle on the side.

Fermenting vegetables is not complicated, but you'll want to have the proper containers, covers, and weights ready before you begin. For containers, keep the following in mind:

- A one-gallon container is needed for each five pounds of fresh vegetables. Therefore, a 5-gallon stone crock is of ideal size for fermenting about 25 pounds of fresh cabbage or cucumbers.
- Food-grade plastic and glass containers are excellent substitutes for stone crocks. Other 1- to 3-gallon non-food-grade plastic containers may be used if lined inside with a clean food-grade plastic bag. **Caution: Be certain that foods contact only food-grade plastics. Do not use garbage bags or trash liners.**
- Fermenting sauerkraut in quart and half-gallon mason jars is an acceptable practice, but may result in more spoilage losses.

Some vegetables, like cabbage and cucumbers, need to be kept 1 to 2 inches under brine while fermenting. If you find them floating to top of the container, here are some suggestions:

- After adding prepared vegetables and brine, insert a suitably sized dinner plate or glass pie plate inside the fermentation container. The plate must be slightly smaller than the container opening, yet large enough to cover most of the shredded cabbage or cucumbers.

- To keep the plate under the brine, weigh it down with two to three sealed quart jars filled with water. Covering the container opening with a clean, heavy bath towel helps to prevent contamination from insects and molds while the vegetables are fermenting.
- Fine quality fermented vegetables are also obtained when the plate is weighed down with a very large, clean, plastic bag filled with three quarts of water containing 4½ tablespoons of salt. Be sure to seal the plastic bag. Freezer bags sold for packaging turkeys are suitable for use with five-gallon containers.

Be sure to wash the fermentation container, plate, and jars in hot, sudsy water, and rinse well with very hot water before use.

Regular dill pickles and sauerkraut are fermented and cured for about three weeks. Refrigerator dills are fermented for about one week. During curing, colors and flavors change and acidity increases. Fresh-pack or quick-process pickles are not fermented; some are brined several hours or overnight, then drained and covered with vinegar and seasonings. Fruit pickles usually are prepared by heating fruit in a seasoned syrup acidified with either lemon juice or vinegar. Relishes are made from chopped fruits and vegetables that are cooked with seasonings and vinegar.

Be sure to remove and discard a 1/16-inch slice from the blossom end of fresh cucumbers. Blossoms may contain an enzyme which causes excessive softening of pickles.

Caution: The level of acidity in a pickled product is as important to its safety as it is to taste and texture.

- **Do not alter vinegar, food, or water proportions in a recipe or use a vinegar with unknown acidity.**
- **Use only recipes with tested proportions of ingredients.**
- **There must be a minimum, uniform level of acid throughout the mixed product to prevent the growth of botulinum bacteria.**

Ingredients

Select fresh, firm fruits or vegetables free of spoilage. Measure or weigh amounts carefully, because the proportion of fresh food to other ingredients will affect flavor and, in many instances, safety.

Use canning or pickling salt. Noncaking material added to other salts may make the brine cloudy. Since flake salt varies in density, it is not recommended for making pickled and fermented foods. White granulated and brown sugars are most often used. Corn syrup and honey, unless called for in reliable recipes, may produce undesirable flavors. White distilled and cider vinegars of 5 percent acidity (50 grain) are recommended. White vinegar is usually preferred when light color is desirable, as is the case with fruits and cauliflower.

Pickles with reduced salt content

In the making of fresh-pack pickles, cucumbers are acidified quickly with vinegar. Use only tested recipes formulated to produce the proper acidity. While these pickles may be prepared safely with reduced or no salt, their quality may be noticeably lower. Both texture and flavor may be slightly, but noticeably, different than expected. You may wish to make small quantities first to determine if you like them.

However, the salt used in making fermented sauerkraut and brined pickles not only provides characteristic flavor but also is vital to safety and texture. In fermented foods, salt favors the growth of desirable bacteria while inhibiting the growth of others. **Caution: Do not attempt to make sauerkraut or fermented pickles by cutting back on the salt required.**

Preventing spoilage

Pickle products are subject to spoilage from microorganisms, particularly yeasts and molds, as well as enzymes that may affect flavor, color, and texture. Processing the pickles in a boiling-water canner will prevent both of these problems. Standard canning jars and self-sealing lids are recommended. Processing times and procedures will vary according to food acidity and the size of food pieces.

Dill Pickles

Feel free to alter the spices in this recipe, but stick to the same proportion of cucumbers, vinegar, and water. Check the label of your vinegar to be sure it contains 5 percent acetic acid. Fully fermented pickles may be stored in the original container for about four to six months, provided they are refrigerated and surface scum and molds are removed regularly, but canning is a better way to store fully fermented pickles.

Ingredients

Use the following quantities for each gallon capacity of your container:

- 4 lbs. of 4-inch pickling cucumbers
- 2 tbsps dill seed or 4 to 5 heads fresh or dry dill weed
- ½ cup salt
- ¼ cup vinegar (5 percent acetic acid)
- 8 cups water
  And one or more of the following ingredients:
- 2 cloves garlic (optional)
- 2 dried red peppers (optional)
- 2 tsps whole mixed pickling spices (optional)

Directions

1. Wash cucumbers. Cut 1/16-inch slice off blossom end and discard. Leave ¼ inch of stem attached. Place half of dill and spices on bottom of a clean, suitable container.
2. Add cucumbers, remaining dill, and spices. Dissolve salt in vinegar and water and pour over cucumbers. Add suitable cover and weight. Store where temperature is between 70 and 75°F for about 3 to 4 weeks while fermenting. Temperatures of 55 to 65°F are acceptable, but the fermentation will take 5 to 6 weeks. Avoid temperatures above 80°F, or pickles will become too soft during fermentation. Fermenting pickles cure slowly. Check the container several times a week and promptly remove surface scum or mold. **Caution: If the pickles become soft, slimy, or develop a disagreeable odor, discard them.**
3. Once fully fermented, pour the brine into a pan, heat slowly to a boil, and simmer 5 minutes. Filter brine through paper coffee filters to reduce cloudiness, if desired. Fill jars with pickles and hot brine, leaving ½-inch headspace. Adjust lids and process in a boiling water canner, or use the low-temperature pasteurization treatment described below.

Low-temperature Pasteurization Treatment

The following treatment results in a better product texture but must be carefully managed to avoid possible spoilage.

1. Place jars in a canner filled halfway with warm (120 to 140°F) water. Then, add hot water to a level 1 inch above jars.
2. Heat the water enough to maintain 180 to 185°F water temperature for 30 minutes. Check with a candy or jelly thermometer to be certain that the water temperature is at least 180°F during the entire 30 minutes. Temperatures higher than 185°F may cause unnecessary softening of pickles.

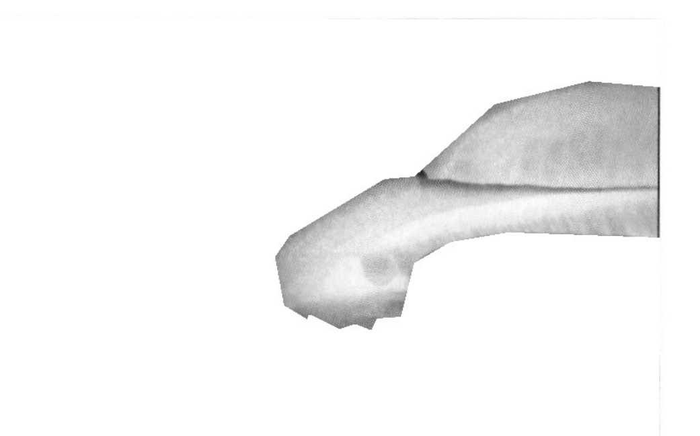

Sauerkraut

For the best sauerkraut, use firm heads of fresh cabbage. Shred cabbage and start kraut between twenty-four and forty-eight hours after harvest. This recipe yields about 9 quarts.

Ingredients

- 25 lbs. cabbage
- ¾ cup canning or pickling salt

Directions

1. Work with about 5 pounds of cabbage at a time. Discard outer leaves. Rinse heads under cold running water and drain. Cut heads in quarters and remove cores. Shred or slice to the thickness of a quarter.
2. Put cabbage in a suitable fermentation container, and add 3 tablespoons of salt. Mix thoroughly, using clean hands. Pack firmly until salt draws juices from cabbage.
3. Repeat shredding, salting, and packing until all cabbage is in the container. Be sure it is deep enough so that its rim is at least 4 or 5 inches above the cabbage. If juice does not cover cabbage, add boiled and cooled brine (1½ tablespoons of salt per quart of water).
4. Add plate and weights; cover container with a clean bath towel. Store at 70 to 75°F while fermenting. At temperatures between 70 and 75°F, kraut will be fully fermented in about 3 to 4 weeks; at 60° to 65°F, fermentation may take 5 to 6 weeks. At temperatures lower than 60°F, kraut may not ferment. Above 75°F, kraut may become soft.

Note: If you weigh the cabbage down with a brine-filled bag, do not disturb the crock until normal fermentation is completed (when bubbling ceases). If you use jars as weight, you will have to check the kraut two to three times each week and remove scum if it forms. Fully fermented kraut may be kept tightly covered in the refrigerator for several months or it may be canned as follows:

Hot pack—Bring kraut and liquid slowly to a boil in a large kettle, stirring frequently. Remove from heat and fill jars rather firmly with kraut and juices, leaving ½-inch headspace.

Raw pack—Fill jars firmly with kraut and cover with juices, leaving ½-inch headspace.

5. Adjust lids and process.

Pickled Three-bean Salad

This is a great side dish to bring to a summer picnic or potluck. Feel free to add or adjust spices to your taste. This recipe yields about 5 to 6 half-pints.

Ingredients

- 1½ cups cut and blanched green or yellow beans (prepared as below)
- 1½ cups canned, drained red kidney beans
- 1 cup canned, drained garbanzo beans

PROCESS TIMES FOR DILL PICKLES IN A BOILING-WATER CANNER*

		Process Time at Altitudes of:		
Style of Pack	Jar Size	0–1,000 ft	1,001–6,000 ft	Above 6,000 ft
Raw	Pints	10 minutes	15 minutes	20 minutes
	Quarts	15 minutes	20 minutes	25 minutes

*After the process is complete, turn off the heat and remove the canner lid. Wait five minutes before removing jars.

PROCESS TIMES FOR SAUERKRAUT IN A BOILING-WATER CANNER*

		Process Time at Altitudes of:			
Style of Pack	Jar Size	0–1,000 ft	1,001-3,000 ft	3,001-6,000 ft	Above 6,000 ft
Hot	Pints	10 minutes	15 minutes	15 minutes	20 minutes
	Quarts	15 minutes	20 minutes	20 minutes	25 minutes
Raw	Pints	20 minutes	25 minutes	30 minutes	35 minutes
	Quarts	25 minutes	30 minutes	35 minutes	40 minutes

*After the process is complete, turn off the heat and remove the canner lid. Wait five minutes before removing jars.

PROCESS TIMES FOR PICKLED THREE-BEAN SALAD IN A BOILING WATER CANNER*

		Process Time at Altitudes of:		
Style of Pack	Jar Size	0–1,000 ft	1,001–6,000 ft	Above 6,000 ft
Hot	Half-pints or Pints	15 minutes	20 minutes	25 minutes

*After the process is complete, turn off the heat and remove the canner lid. Wait five minutes before removing jars.

- ½ cup peeled and thinly sliced onion (about 1 medium onion)
- ½ cup trimmed and thinly sliced celery (1½ medium stalks)
- ½ cup sliced green peppers (½ medium pepper)
- ½ cup white vinegar (5 percent acetic acid)
- ¼ cup bottled lemon juice
- ¾ cup sugar
- 1¼ cups water
- ¼ cup oil
- ½ tsp canning or pickling salt

Directions

1. Wash and snap off ends of fresh beans. Cut or snap into 1- to 2-inch pieces. Blanch 3 minutes and cool immediately. Rinse kidney beans with tap water and drain again. Prepare and measure all other vegetables.
2. Combine vinegar, lemon juice, sugar, and water and bring to a boil. Remove from heat. Add oil and salt and mix well. Add beans, onions, celery, and green pepper to solution and bring to a simmer.
3. Marinate 12 to 14 hours in refrigerator, then heat entire mixture to a boil. Fill clean jars with solids. Add hot liquid, leaving ½-inch headspace. Adjust lids and process.

Pickled Horseradish Sauce

Select horseradish roots that are firm and have no mold, soft spots, or green spots. Avoid roots that have begun to sprout. The pungency of fresh horseradish fades within one to two months, even when refrigerated, so make only small quantities at a time. This recipe yields about two half-pints.

Ingredients

- 2 cups (¾ lb.) freshly grated horseradish
- 1 cup white vinegar (5 percent acetic acid)
- ½ tsp canning or pickling salt
- ¼ tsp powdered ascorbic acid

Directions

1. Wash horseradish roots thoroughly and peel off brown outer skin. Grate the peeled roots in a food processor or cut them into small cubes and put through a food grinder.
2. Combine ingredients and fill into sterile jars, leaving ¼-inch headspace. Seal jars tightly and store in a refrigerator.

Marinated Peppers

Any combination of bell, Hungarian, banana, or jalapeño peppers can be used in this recipe. Use more jalapeño peppers if you want your mix to be hot, but remember to wear rubber or plastic gloves while handling them or wash hands thoroughly with soap and water before touching your face. This recipe yields about 9 half-pints.

Ingredients

- 4 lbs. firm peppers
- 1 cup bottled lemon juice
- 2 cups white vinegar (5 percent acetic acid)
- 1 tbsp oregano leaves
- 1 cup olive or salad oil
- ½ cup chopped onions
- 2 tbsps prepared horseradish (optional)
- 2 cloves garlic, quartered (optional)
- 2¼ tsps salt (optional)

Directions

1. Select your favorite pepper. Peppers may be left whole or quartered. Wash, slash two to four slits in each pepper, and blanch in boiling water or blister in order to peel tough-skinned hot peppers. Blister peppers using one of the following methods:

Oven or broiler method—Place peppers in a hot oven (400°F) or broiler for 6 to 8 minutes or until skins blister.

Range-top method—Cover hot burner, either gas or electric, with heavy wire mesh. Place peppers on burner for several minutes until skins blister.

2. Allow peppers to cool. Place in pan and cover with a damp cloth. This will make peeling the peppers easier. After several minutes of cooling, peel each pepper. Flatten whole peppers.
3. Mix all remaining ingredients except garlic and salt in a saucepan and heat to boiling. Place ¼ garlic clove (optional) and ¼ teaspoon salt in each half-pint or ½ teaspoon per pint. Fill jars with peppers, and add hot, well-mixed oil/pickling solution over peppers, leaving ½-inch headspace. Adjust lids and process.

Piccalilli

Piccalilli is a nice accompaniment to roasted or braised meats and is common in British and Indian meals. It can also be mixed with mayonnaise or crème fraîche as the basis of a French remoulade. This recipe yields 9 half-pints.

PROCESS TIMES FOR MARINATED PEPPERS IN A BOILING-WATER CANNER*

		Process Time at Altitudes of:			
Style of Pack	Jar Size	0–1,000 ft	1,001–3,000 ft	3,001–6,000 ft	Above 6,000 ft
Raw	Half-pints and Pints	15 minutes	20 minutes	20 minutes	25 minutes

*After the process is complete, turn off the heat and remove the canner lid. Wait five minutes before removing jars.

Ingredients

- 6 cups chopped green tomatoes
- 1½ cups chopped sweet red peppers
- 1½ cups chopped green peppers
- 2¼ cups chopped onions
- 7½ cups chopped cabbage
- ½ cup canning or pickling salt
- 3 tbsps whole mixed pickling spice
- 4½ cups vinegar (5 percent acetic acid)
- 3 cups brown sugar

Directions

1. Wash, chop, and combine vegetables with salt. Cover with hot water and let stand 12 hours. Drain and press in a clean white cloth to remove all possible liquid.
2. Tie spices loosely in a spice bag and add to combined vinegar and brown sugar and heat to a boil in a saucepan. Add vegetables and boil gently 30 minutes or until the volume of the mixture is reduced by one-half. Remove spice bag.
3. Fill hot sterile jars with hot mixture, leaving ½-inch headspace. Adjust lids and process.

Bread-and-Butter Pickles

These slightly sweet, spiced pickles will add flavor and crunch to any sandwich. If desired, slender (1 to 1½ inches in diameter) zucchini or yellow summer squash can be substituted for cucumbers. After processing and cooling, jars should be stored four to five weeks to develop ideal flavor. This recipe yields about 8 pints.

Ingredients

- 6 lbs. of 4- to 5-inch pickling cucumbers
- 8 cups thinly sliced onions (about 3 pounds)
- ½ cup canning or pickling salt
- 4 cups vinegar (5 percent acetic acid)
- 4½ cups sugar
- 2 tbsps mustard seed
- 1½ tbsps celery seed
- 1 tbsp ground turmeric
- 1 cup pickling lime (optional—for use in variation below for making firmer pickles)

Directions

1. Wash cucumbers. Cut 1/16 inch off blossom end and discard. Cut into 3/16-inch slices. Combine cucumbers and onions in a large bowl. Add salt. Cover with 2 inches crushed or cubed ice. Refrigerate 3 to 4 hours, adding more ice as needed.
2. Combine remaining ingredients in a large pot. Boil ten minutes. Drain cucumbers and onions, add to pot, and slowly reheat to boiling. Fill jars with slices and cooking syrup, leaving ½-inch headspace.

Variation for firmer pickles: Wash cucumbers. Cut 1/16 inch off blossom end and discard. Cut into 3/16-inch slices. Mix 1 cup pickling lime and ½ cup salt to 1 gallon water in a 2- to 3-gallon crock or enamelware container. Avoid inhaling lime dust while mixing the lime-water solution. Soak cucumber slices in lime water for twelve to twenty-four hours, stirring occasionally. Remove from lime solution, rinse, and resoak one hour in fresh cold water. Repeat the rinsing and soaking steps two more times. Handle carefully, as slices will be brittle. Drain well.

PROCESS TIMES FOR PICCALILLI IN A BOILING-WATER CANNER*

		Process Time at Altitudes of:		
Style of Pack	Jar Size	0–1,000 ft	1,001–6,000 ft	Above 6,000 ft
Hot	Half-pints or Pints	5 minutes	10 minutes	15 minutes

*After the process is complete, turn off the heat and remove the canner lid. Wait five minutes before removing jars.

PROCESS TIMES FOR BREAD-AND-BUTTER PICKLES IN A BOILING-WATER CANNER*

		Process Time at Altitudes of:		
Style of Pack	**Jar Size**	**0–1,000 ft**	**1,001–6,000 ft**	**Above 6,000 ft**
Hot	Pints or Quarts	10 minutes	15 minutes	20 minutes
*After the process is complete, turn off the heat and remove the canner lid. Wait five minutes before removing jars.				

3. Adjust lids and process in boiling-water canner, or use the low-temperature pasteurization treatment described below.

Low-temperature Pasteurization Treatment

The following treatment results in a better product texture but must be carefully managed to avoid possible spoilage.

1. Place jars in a canner filled halfway with warm (120 to 140°F) water. Then, add hot water to a level 1 inch above jars.
2. Heat the water enough to maintain 180 to 185°F water temperature for 30 minutes. Check with a candy or jelly thermometer to be certain that the water temperature is at least 180°F during the entire 30 minutes. Temperatures higher than 185°F may cause unnecessary softening of pickles.

Quick Fresh-pack Dill Pickles

For best results, pickle cucumbers within twenty-four hours of harvesting, or immediately after purchasing. This recipe yields seven to nine pints.

Ingredients

- 8 lbs. of 3- to 5-inch pickling cucumbers
- 2 gallons water
- 1¼ to 1½ cups canning or pickling salt
- 1½ qts vinegar (5 percent acetic acid)
- ¼ cup sugar
- 2 to 2¼ quarts water
- 2 tbsps whole mixed pickling spice
- 3 to 5 tbsps whole mustard seed (1 to 2 tsps per pint jar)
- 14 to 21 heads of fresh dill (1½ to 3 heads per pint jar) *or*
- 4½ to 7 tbsps dill seed (1½ tsps to 1 tbsp per pint jar)

PROCESS TIMES FOR QUICK FRESH-PACK DILL PICKLES IN A BOILING-WATER CANNER*

		Process Time at Altitudes of:		
Style of Pack	**Jar Size**	**0–1,000 ft**	**1,001–6,000 ft**	**Above 6,000 ft**
Raw	Pints	10 minutes	15 minutes	20 minutes
	Quarts	15 minutes	20 minutes	25 minutes
*After the process is complete, turn off the heat and remove the canner lid. Wait five minutes before removing jars.				

PROCESS TIMES FOR PICKLE RELISH IN A BOILING-WATER CANNER*

		Process Time at Altitudes of:		
Style of Pack	**Jar Size**	**0–1,000 ft**	**1,001–6,000 ft**	**Above 6,000 ft**
Hot	Half-pints or Pints	10 minutes	15 minutes	20 minutes
*After the process is complete, turn off the heat and remove the canner lid. Wait five minutes before removing jars.				

Directions

1. Wash cucumbers. Cut 1/16-inch slice off blossom end and discard, but leave ¼-inch of stem attached. Dissolve ¾ cup salt in 2 gallons water. Pour over cucumbers and let stand twelve hours. Drain.
2. Combine vinegar, ½ cup salt, sugar, and 2 quarts water. Add mixed pickling spices tied in a clean white cloth. Heat to boiling. Fill jars with cucumbers. Add 1 tsp mustard seed and 1½ heads fresh dill per pint.
3. Cover with boiling pickling solution, leaving ½-inch headspace. Adjust lids and process.

Pickle Relish

A food processor will make quick work of chopping the vegetables in this recipe. Yields about 9 pints.

Ingredients

- 3 qts chopped cucumbers
- 3 cups each of chopped sweet green and red peppers
- 1 cup chopped onions
- ¾ cup canning or pickling salt
- 4 cups ice
- 8 cups water
- 4 tsp each of mustard seed, turmeric, whole allspice, and whole cloves
- 2 cups sugar
- 6 cups white vinegar (5 percent acetic acid)

Directions

1. Add cucumbers, peppers, onions, salt, and ice to water and let stand 4 hours. Drain and re-cover vegetables with fresh ice water for another hour. Drain again.
2. Combine spices in a spice or cheesecloth bag. Add spices to sugar and vinegar. Heat to boiling and pour mixture over vegetables. Cover and refrigerate 24 hours.
3. Heat mixture to boiling and fill hot into clean jars, leaving ½-inch headspace. Adjust lids and process.

Quick Sweet Pickles

Quick and simple to prepare, these are the sweet pickles to make when you're short on time. After processing and cooling, jars should be stored four to five weeks to develop ideal flavor. If desired, add two slices of raw whole onion to each jar before filling with cucumbers. This recipe yields about 7 to 9 pints.

Ingredients

- 8 lbs of 3- to 4-inch pickling cucumbers
- ⅓ cup canning or pickling salt
- 4½ cups sugar
- 3½ cups vinegar (5 percent acetic acid)
- 2 tsps celery seed
- 1 tbsp whole allspice
- 2 tbsps mustard seed
- 1 cup pickling lime (optional)

Directions

1. Wash cucumbers. Cut 1/16 inch off blossom end and discard, but leave ¼ inch of stem attached. Slice or cut in strips, if desired.
2. Place in bowl and sprinkle with salt. Cover with 2 inches of crushed or cubed ice. Refrigerate 3 to 4 hours. Add more ice as needed. Drain well.
3. Combine sugar, vinegar, celery seed, allspice, and mustard seed in 6-quart kettle. Heat to boiling.

Hot pack—Add cucumbers and heat slowly until vinegar solution returns to boil. Stir occasionally to make sure mixture heats evenly. Fill sterile jars, leaving ½-inch headspace.

Raw pack—Fill jars, leaving ½-inch headspace.

4. Add hot pickling syrup, leaving ½-inch headspace. Adjust lids and process.

Variation for firmer pickles: Wash cucumbers. Cut 1/16 inch off blossom end and discard, but leave ¼ inch of stem attached. Slice or strip cucumbers. Mix 1 cup pickling lime and 1/3 cup salt with 1 gallon water in a 2- to 3-gallon crock or enamelware container. **Caution: Avoid inhaling lime dust while mixing the lime-water solution.** Soak cucumber slices or strips in lime-water solution for twelve to twenty-four hours, stirring occasionally. Remove from lime solution, rinse, and soak 1 hour in fresh cold water. Repeat the rinsing and soaking two more times. Handle carefully, because slices or strips will be brittle. Drain well.

Reduced-sodium Sliced Sweet Pickles

Whole allspice can be tricky to find. If it's not available at your local grocery store, it can be ordered at www.spicebarn.com or at www.gourmetsleuth.com. This recipe yields about 4 to 5 pints.

Ingredients

- 4 lbs (3- to 4-inch) pickling cucumbers

Canning syrup:

- 1⅔ cups distilled white vinegar (5 percent acetic acid)
- 3 cups sugar
- 1 tbsp whole allspice
- 2¼ tsps celery seed

Brining solution:

- 1 qt distilled white vinegar (5 percent acetic acid)
- 1 tbsp canning or pickling salt
- 1 tbsp mustard seed
- ½ cup sugar

Directions

1. Wash cucumbers and cut 1/2 inch off blossom end, and discard. Cut cucumbers into ¼-inch slices. Combine all ingredients for canning syrup in a saucepan and bring to boiling. Keep syrup hot until used.

PROCESS TIMES FOR QUICK SWEET PICKLES IN A BOILING-WATER CANNER*

		Process Time at Altitudes of:		
Style of Pack	Jar Size	0–1,000 ft	1,001-6,000 ft	Above 6,000 ft
Hot	Pints or Quarts	5 minutes	10 minutes	15 minutes
Raw	Pints	10 minutes	15 minutes	20 minutes
	Quarts	15 minutes	20 minutes	25 minutes

*After the process is complete, turn off the heat and remove the canner lid. Wait five minutes before removing jars.

PROCESS TIMES FOR REDUCED-SODIUM SLICED SWEET PICKLES IN A BOILING-WATER CANNER*

		Process Time at Altitudes of:		
Style of Pack	Jar Size	0–1,000 ft	1,001–6,000 ft	Above 6,000 ft
Hot	Pints	10 minutes	15 minutes	20 minutes

*After the process is complete, turn off the heat and remove the canner lid. Wait five minutes before removing jars.

2. In a large kettle, mix the ingredients for the brining solution. Add the cut cucumbers, cover, and simmer until the cucumbers change color from bright to dull green (about 5 to 7 minutes). Drain the cucumber slices.
3. Fill jars, and cover with hot canning syrup leaving ½-inch headspace. Adjust lids and process.

Tomatoes

Canned tomatoes should be a staple in every cook's pantry. They are easy to prepare and, when made with garden-fresh produce, make ordinary soups, pizza, or pastas into five-star meals. Be sure to select only disease-free, preferably vine-ripened, firm fruit. Do not can tomatoes from dead or frost-killed vines.

Green tomatoes are more acidic than ripened fruit and can be canned safely with the following recommendations.

- To ensure safe acidity in whole, crushed, or juiced tomatoes, add two tablespoons of bottled lemon juice or ½ teaspoon of citric acid per quart of tomatoes. For pints, use one tablespoon bottled lemon juice or ¼ teaspoon citric acid.
- Acid can be added directly to the jars before filling with product. Add sugar to offset acid taste, if desired. Four tablespoons of 5 percent acidity vinegar per quart may be used instead of lemon juice or citric acid. However, vinegar may cause undesirable flavor changes.
- Using a pressure canner will result in higher quality and more nutritious canned tomato products. If your pressure canner cannot be operated above 15 PSI, select a process time at a lower pressure.

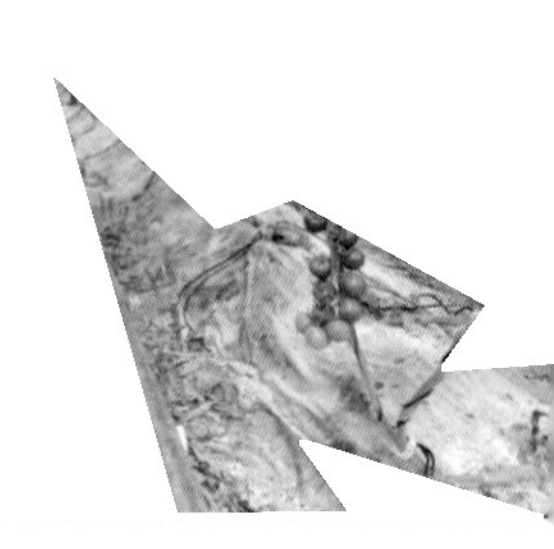

Tomato Juice

Tomato juice is a good source of vitamin A and C and is tasty on its own or in a cocktail. It's also the secret ingredient in some very delicious cakes. If desired, add carrots, celery, and onions, or toss in a few jalapeños for a little kick.

Directions

1. Wash tomatoes, remove stems, and trim off bruised or discolored portions. To prevent juice from separating, quickly cut about 1 pound of fruit into quarters

QUANTITY

- An average of 23 pounds of tomatoes is needed per canner load of 7 quarts, or an average of 14 pounds per canner load of 9 pints.
- A bushel weighs 53 pounds and yields 15 to 18 quarts of juice—an average of 3¼ pounds per quart.

PROCESS TIMES FOR TOMATO JUICE IN A BOILING-WATER CANNER*

		Process Time at Altitudes of:			
Style of Pack	Jar Size	0–1,000 ft	1,001–3,000 ft	3,001–6,000 ft	Above 6,000 ft
Hot	Pints	35 minutes	40 minutes	45 minutes	50 minutes
	Quarts	40 minutes	45 minutes	50 minutes	55 minutes

*After the process is complete, turn off the heat and remove the canner lid. Wait five minutes before removing jars.

PROCESS TIMES FOR TOMATO JUICE IN A DIAL-GAUGE PRESSURE CANNER*

			Canner Gauge Pressure (PSI) at Altitudes of:			
Style of Pack	Jar Size	Process Time	0–2,000 ft	2,001–4,000 ft	4,001–6,000 ft	6,001–8,000 ft
Hot	Pints	20 minutes	6 lbs	7 lbs	8 lbs	9 lbs
	Quarts	15 minutes	11 lbs	12 lbs	13 lbs	14 lbs

*After the canner is completely depressurized, remove the weight from the vent port or open the petcock. Wait ten minutes; then unfasten the lid and remove it carefully. Lift the lid with the underside away from you so that the steam coming out of the canner does not burn your face.

PROCESS TIMES FOR TOMATO JUICE IN A WEIGHTED-GAUGE PRESSURE CANNER*

			Canner Gauge Pressure (PSI) at Altitudes of:	
Style of Pack	Jar Size	Process Time	0–1,000 ft	Above 1,000 ft
Hot	Pints	20 minutes	5 lbs	10 lbs
	Quarts	15 minutes	10 lbs	15 lbs

*After the canner is completely depressurized, remove the weight from the vent port or open the petcock. Wait ten minutes; then unfasten the lid and remove it carefully. Lift the lid with the underside away from you so that the steam coming out of the canner does not burn your face.

and put directly into saucepan. Heat immediately to boiling while crushing.

2. Continue to slowly add and crush freshly cut tomato quarters to the boiling mixture. Make sure the mixture boils constantly and vigorously while you add the remaining tomatoes. Simmer 5 minutes after you add all pieces.
3. Press heated juice through a sieve or food mill to remove skins and seeds. Add bottled lemon juice or citric acid to jars (see page 117). Heat juice again to boiling.
4. Add 1 teaspoon of salt per quart to the jars, if desired. Fill jars with hot tomato juice, leaving ½-inch headspace. Adjust lids and process.

Crushed Tomatoes with No Added Liquid

Crushed tomatoes are great for use in soups, stews, thick sauces, and casseroles. Simmer crushed tomatoes with kidney beans, chili powder, sautéed onions, and garlic to make an easy pot of chili.

Directions

1. Wash tomatoes and dip in boiling water for 30 to 60 seconds or until skins split. Then dip in cold water, slip off skins, and remove cores. Trim off any bruised or discolored portions and quarter.
2. Heat 1/6 of the quarters quickly in a large pot, crushing them with a wooden mallet or spoon as they are added to

PROCESS TIMES FOR CRUSHED TOMATOES IN A DIAL-GAUGE PRESSURE CANNER*

			Canner Gauge Pressure (PSI) at Altitudes of:			
Style of Pack	Jar Size	Process Time	0–2,000 ft	2,001–4,000 ft	4,001–6,000 ft	6,001–8,000 ft
Hot	Pints	20 minutes	6 lbs	7 lbs	8 lbs	9 lbs
	Quarts	15 minutes	11 lbs	12 lbs	13 lbs	14 lbs

*After the canner is completely depressurized, remove the weight from the vent port or open the petcock. Wait ten minutes; then unfasten the lid and remove it carefully. Lift the lid with the underside away from you so that the steam coming out of the canner does not burn your face.

PROCESS TIMES FOR CRUSHED TOMATOES IN A WEIGHTED-GAUGE PRESSURE CANNER*

			Canner Gauge Pressure (PSI) at Altitudes of:	
Style of Pack	Jar Size	Process Time	0–1,000 ft	Above 1,000 ft
Hot	Pints	20 minutes	5 lbs	10 lbs
	Quarts	15 minutes	10 lbs	15 lbs

*After the canner is completely depressurized, remove the weight from the vent port or open the petcock. Wait ten minutes; then unfasten the lid and remove it carefully. Lift the lid with the underside away from you so that the steam coming out of the canner does not burn your face.

PROCESS TIMES FOR CRUSHED TOMATOES IN A BOILING-WATER CANNER*

		Process Time at Altitudes of:			
Style of Pack	Jar Size	0–1,000 ft	1,001–3,000 ft	3,001–6,000 ft	Above 6,000 ft
Hot	Pints	35 minutes	40 minutes	45 minutes	50 minutes
	Quarts	45 minutes	50 minutes	55 minutes	60 minutes

*After the process is complete, turn off the heat and remove the canner lid. Wait five minutes before removing jars.

QUANTITY

- An average of 22 pounds of tomatoes is needed per canner load of 7 quarts.
- An average of 14 fresh pounds is needed per canner load of 9 pints.
- A bushel weighs 53 pounds and yields 17 to 20 quarts of crushed tomatoes—an average of 2¾ pounds per quart.

the pot. This will exude juice. Continue heating the tomatoes, stirring to prevent burning.

3. Once the tomatoes are boiling, gradually add remaining quartered tomatoes, stirring constantly. These remaining tomatoes do not need to be crushed; they will soften with heating and stirring. Continue until all tomatoes are added. Then boil gently 5 minutes.
4. Add bottled lemon juice or citric acid to jars (see page 117). Add 1 teaspoon of salt per quart to the jars, if desired. Fill jars immediately with hot tomatoes, leaving ½-inch headspace. Adjust lids and process.

QUANTITY

For thin sauce:

- An average of 35 pounds of tomatoes is needed per canner load of 7 quarts.
- An average of 21 pounds is needed per canner load of 9 pints.
- A bushel weighs 53 pounds and yields 10 to 12 quarts of sauce—an average of 5 pounds per quart.

For thick sauce:

- An average of 46 pounds is needed per canner load of 7 quarts.
- An average of 28 pounds is needed per canner load of 9 pints.
- A bushel weighs 53 pounds and yields 7 to 9 quarts of sauce—an average of 6½ pounds per quart.

PROCESS TIMES FOR TOMATO SAUCE IN A BOILING-WATER CANNER*

		Process Time at Altitudes of:			
Style of Pack	Jar Size	0–1,000 ft	1,001–3,000 ft	3,001–6,000 ft	Above 6,000 ft
Hot	Pints	35 minutes	40 minutes	45 minutes	50 minutes
	Quarts	40 minutes	45 minutes	50 minutes	55 minutes

*After the process is complete, turn off the heat and remove the canner lid. Wait five minutes before removing jars.

PROCESS TIMES FOR TOMATO SAUCE IN A DIAL-GAUGE PRESSURE CANNER*

			Canner Gauge Pressure (PSI) at Altitudes of:			
Style of Pack	Jar Size	Process Time	0–2,000 ft	2,001–4,000 ft	4,001–6,000 ft	6,001–8,000 ft
Hot	Pints	20 minutes	6 lbs	7 lbs	8 lbs	9 lbs
	Quarts	15 minutes	11 lbs	12 lbs	13 lbs	14 lbs

*After the canner is completely depressurized, remove the weight from the vent port or open the petcock. Wait ten minutes; then unfasten the lid and remove it carefully. Lift the lid with the underside away from you so that the steam coming out of the canner does not burn your face.

PROCESS TIMES FOR TOMATO SAUCE IN A WEIGHTED-GAUGE PRESSURE CANNER*

			Canner Gauge Pressure (PSI) at Altitudes of:	
Style of Pack	Jar Size	Process Time	0–1,000 ft	Above 1,000 ft
Hot	Pints	20 minutes	5 lbs	10 lbs
	Quarts	15 minutes	10 lbs	15 lbs

*After the canner is completely depressurized, remove the weight from the vent port or open the petcock. Wait ten minutes; then unfasten the lid and remove it carefully. Lift the lid with the underside away from you so that the steam coming out of the canner does not burn your face.

Tomato Sauce

This plain tomato sauce can be spiced up before using in soups or in pink or red sauces. The thicker you want your sauce, the more tomatoes you'll need.

Directions

1. Prepare and press as for making tomato juice (see page 147). Simmer in a large saucepan until sauce reaches desired consistency. Boil until volume is reduced by about one-third for thin sauce, or by one-half for thick sauce.
2. Add bottled lemon juice or citric acid to jars (see page 117). Add 1 teaspoon of salt per quart to the jars, if desired. Fill jars, leaving ¼-inch headspace. Adjust lids and process.

Tomatoes, Whole or Halved, Packed in Water

Whole or halved tomatoes are used for scalloped tomatoes, savory pies (baked in a pastry crust with parmesan cheese, mayonnaise, and seasonings), or stewed tomatoes.

PROCESS TIMES FOR WATER-PACKED WHOLE TOMATOES IN A BOILING-WATER CANNER*

		Process Time at Altitudes of:			
Style of Pack	Jar Size	0–1,000 ft	1,001–3,000 ft	3,001–6,000 ft	Above 6,000 ft
Hot or Raw	Pints	40 minutes	45 minutes	50 minutes	55 minutes
	Quarts	45 minutes	50 minutes	55 minutes	60 minutes

*After the process is complete, turn off the heat and remove the canner lid. Wait five minutes before removing jars.

PROCESS TIMES FOR WATER-PACKED WHOLE TOMATOES IN A DIAL-GAUGE PRESSURE CANNER*

			Canner Gauge Pressure (PSI) at Altitudes of:			
Style of Pack	Jar Size	Process Time	0–2,000 ft	2,001–4,000 ft	4,001–6,000 ft	6,001–8,000 ft
Hot or Raw	Pints	15 minutes	6 lbs	7 lbs	8 lbs	9 lbs
	Quarts	10 minutes	11 lbs	12 lbs	13 lbs	14 lbs

*After the canner is completely depressurized, remove the weight from the vent port or open the petcock. Wait ten minutes; then unfasten the lid and remove it carefully. Lift the lid with the underside away from you so that the steam coming out of the canner does not burn your face.

PROCESS TIMES FOR WATER-PACKED WHOLE TOMATOES IN A WEIGHTED-GAUGE PRESSURE CANNER*

			Canner Gauge Pressure (PSI) at Altitudes of:	
Style of Pack	Jar Size	Process Time	0–1,000 ft	Above 1,000 ft
Hot or Raw	Pints	15 minutes	5 lbs	10 lbs
	Quarts	10 minutes	10 lbs	15 lbs

*After the canner is completely depressurized, remove the weight from the vent port or open the petcock. Wait ten minutes; then unfasten the lid and remove it carefully. Lift the lid with the underside away from you so that the steam coming out of the canner does not burn your face.

QUANTITY

- An average of 21 pounds of tomatoes is needed per canner load of 7 quarts.
- An average of 13 pounds is needed per canner load of 9 pints.
- A bushel weighs 53 pounds and yields 15 to 21 quarts—an average of 3 pounds per quart.

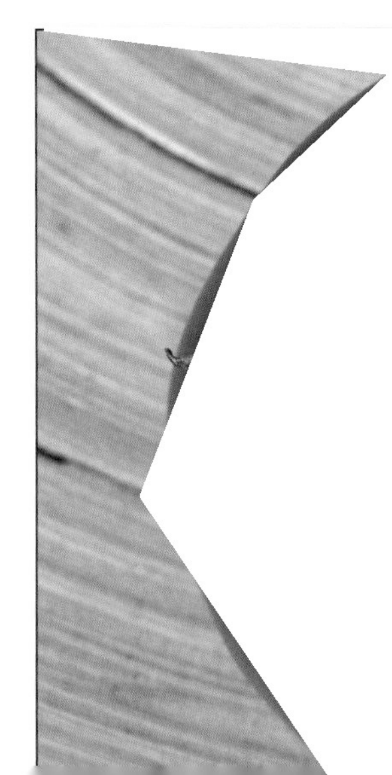

Directions

1. Wash tomatoes. Dip in boiling water for 30 to 60 seconds or until skins split; then dip in cold water. Slip off skins and remove cores. Leave whole or halved.
2. Add bottled lemon juice or citric acid to jars (see page 117). Add 1 teaspoon of salt per quart to the jars, if desired. For hot pack products, add enough water to cover the tomatoes and boil them gently for 5 minutes.
3. Fill jars with hot tomatoes or with raw peeled tomatoes. Add the hot cooking liquid to the hot pack, or hot water for raw pack to cover, leaving ½-inch headspace. Adjust lids and process.

Spaghetti Sauce Without Meat

Homemade spaghetti sauce is like a completely different food than store-bought varieties—it tastes fresher, is more flavorful, and is far more nutritious. Adjust spices to taste, but do not increase proportions of onions, peppers, or mushrooms. This recipe yields about 9 pints.

Ingredients

- 30 lbs. tomatoes
- 1 cup chopped onions
- 5 cloves garlic, minced
- 1 cup chopped celery or green pepper
- 1 lb. fresh mushrooms, sliced (optional)
- 4½ tsps salt
- 2 tbsps oregano
- 4 tbsps minced parsley
- 2 tsps black pepper
- ¼ cup brown sugar
- ¼ cup vegetable oil

Directions

1. Wash tomatoes and dip in boiling water for 30 to 60 seconds or until skins split. Dip in cold water and slip off skins. Remove cores and quarter tomatoes. Boil 20 minutes, uncovered, in large saucepan. Put through food mill or sieve.

PROCESS TIMES FOR SPAGHETTI SAUCE WITHOUT MEAT IN A DIAL-GAUGE PRESSURE CANNER*

			Canner Gauge Pressure (PSI) at Altitudes of:			
Style of Pack	Jar Size	Process Time	0–2,000 ft	2,001–4,000 ft	4,001–6,000 ft	6,001–8,000 ft
Hot	Pints	20 minutes	11 lbs	12 lbs	13 lbs	14 lbs
	Quarts	25 minutes	11 lbs	12 lbs	13 lbs	14 lbs

*After the canner is completely depressurized, remove the weight from the vent port or open the petcock. Wait ten minutes; then unfasten the lid and remove it carefully. Lift the lid with the underside away from you so that the steam coming out of the canner does not burn your face.

PROCESS TIMES FOR SPAGHETTI SAUCE WITHOUT MEAT IN A WEIGHTED-GAUGE PRESSURE CANNER*

			Canner Gauge Pressure (PSI) at Altitudes of:	
Style of Pack	Jar Size	Process Time	0–1,000 ft	Above 1,000 ft
Hot	Pints	20 minutes	10 lbs	15 lbs
	Quarts	25 minutes	10 lbs	15 lbs

*After the canner is completely depressurized, remove the weight from the vent port or open the petcock. Wait ten minutes; then unfasten the lid and remove it carefully. Lift the lid with the underside away from you so that the steam coming out of the canner does not burn your face.

2. Sauté onions, garlic, celery or peppers, and mushrooms (if desired) in vegetable oil until tender. Combine sautéed vegetables and tomatoes and add spices, salt, and sugar. Bring to a boil.
3. Simmer uncovered, until thick enough for serving. Stir frequently to avoid burning. Fill jars, leaving 1-inch headspace. Adjust lids and process.

Tomato Ketchup

Ketchup forms the base of several condiments, including Thousand Island dressing, fry sauce, and barbecue sauce. And, of course, it's an American favorite in its own right. This recipe yields 6 to 7 pints.

Ingredients

- 24 lbs. ripe tomatoes
- 3 cups chopped onions
- ¾ tsp ground red pepper (cayenne)
- 4 tsps whole cloves
- 3 sticks cinnamon, crushed
- 1½ tsp whole allspice
- 3 tbsps celery seeds
- 3 cups cider vinegar (5 percent acetic acid)
- 1½ cups sugar
- ¼ cup salt

Directions

1. Wash tomatoes. Dip in boiling water for 30 to 60 seconds or until skins split. Dip in cold water. Slip off skins and remove cores. Quarter tomatoes into 4-gallon stockpot or a large kettle. Add onions and red pepper. Bring to boil and simmer 20 minutes, uncovered.
2. Combine remaining spices in a spice bag and add to vinegar in a 2-quart saucepan. Bring to boil. Turn off heat and let stand until tomato mixture has been cooked 20 minutes. Then, remove spice bag and combine vinegar and tomato mixture. Boil about 30 minutes.
3. Put boiled mixture through a food mill or sieve. Return to pot. Add sugar and salt, boil gently, and stir frequently until volume is reduced by one-half or until mixture rounds up on spoon without separation. Fill pint jars, leaving 1/8-inch headspace. Adjust lids and process.

PROCESS TIMES FOR TOMATO KETCHUP IN A BOILING-WATER CANNER*

		Process Time at Altitudes of:		
Style of Pack	Jar Size	0–1,000 ft	1,001-6,000 ft	Above 6,000 ft
Hot	Pints	15 minutes	20 minutes	25 minutes

*After the process is complete, turn off the heat and remove the canner lid. Wait five minutes before removing jars.

Chile Salsa (Hot Tomato-Pepper Sauce)

For fantastic nachos, cover corn chips with chile salsa, add shredded Monterey jack or cheddar cheese, bake under broiler for about five minutes, and serve with guacamole and sour cream. Be sure to wear rubber gloves while handling chiles or wash hands thoroughly with soap and water before touching your face. This recipe yields 6 to 8 pints.

Ingredients

- 5 lbs. tomatoes
- 2 lbs. chile peppers
- 1 lb onions
- 1 cup vinegar (5 percent acetic acid)
- 3 tsps salt
- ½ tsp pepper

Directions

1. Wash and dry chiles. Slit each pepper on its side to allow steam to escape. Peel peppers using one of the following methods:

Oven or broiler method: Place chiles in oven (400°F) or broiler for 6 to 8 minutes until skins blister. Cool and slip off skins.

Range-top method: Cover hot burner, either gas or electric, with heavy wire mesh. Place chiles on burner for several minutes until skins blister. Allow peppers to cool. Place in a pan and cover with a damp cloth. This will make peeling the peppers easier. After several minutes, peel each pepper.

2. Discard seeds and chop peppers. Wash tomatoes and dip in boiling water for 30 to 60 seconds or until skins split. Dip in cold water, slip off skins, and remove cores.
3. Coarsely chop tomatoes and combine chopped peppers, onions, and remaining ingredients in a large saucepan. Heat to boil, and simmer ten minutes. Fill jars, leaving ½-inch headspace. Adjust lids and process.

PROCESS TIMES FOR CHILE SALSA IN A BOILING-WATER CANNER*

		Process Time at Altitudes of:		
Style of Pack	Jar Size	0–1,000 ft	1,001–6,000 ft	Above 6,000 ft
Hot	Pints	15 minutes	20 minutes	25 minutes

*After the process is complete, turn off the heat and remove the canner lid. Wait five minutes before removing jars.

Drying and Freezing

Drying

Drying fruits, vegetables, herbs, and even meat is a great way to preserve foods for longer-term storage, especially if your pantry or freezer space is limited. Dried foods take up much less space than their fresh, frozen, or canned counterparts. Drying requires relatively little preparation time, and is simple enough that kids will enjoy helping. Drying with a food dehydrator will ensure the fastest, safest, and best quality results. However, you can also dry produce in the sunshine, in your oven, or strung up over a woodstove.

For more information on food drying, check out *So Easy to Preserve, 5th ed.* from the Cooperative Extension Service, the University of Georgia. Much of the information that follows is adapted from this excellent source.

Drying with a Food Dehydrator

Food dehydrators use electricity to produce heat and have a fan and vents for air circulation. Dehydrators are efficiently designed to dry foods fast at around 140°F. Look for food dehydrators in discount department stores, mail-order catalogs, the small appliance section of a department store, natural food stores, and seed or garden supply catalogs. Costs

vary depending on features. Some models are expandable and additional trays can be purchased later. Twelve square feet of drying space dries about a half-bushel of produce.

Dehydrator Features to Look For

- Double-wall construction of metal or high-grade plastic. Wood is not recommended, because it is a fire hazard and is difficult to clean.
- Enclosed heating elements
- Countertop design
- An enclosed thermostat from 85 to 160°F
- Fan or blower
- Four to ten open mesh trays made of sturdy, lightweight plastic for easy washing
- Underwriters Laboratory (UL) seal of approval
- A one-year guarantee
- Convenient service
- A dial for regulating temperature
- A timer. Often the completed drying time may occur during the night, and a timer turns the dehydrator off to prevent scorching.

Types of Dehydrators

There are two basic designs for dehydrators. One has horizontal air flow and the other has vertical air flow. In units with horizontal flow, the heating element and fan are located on the side of the unit. The major advantages of horizontal flow are: it reduces flavor mixture so several different foods can be dried at one time; all trays receive equal heat penetration; and juices or liquids do not drip down into the heating element. Vertical air flow dehydrators have the heating element and fan located at the base. If different foods are dried, flavors can mix and liquids can drip into the heating element.

Fruit Drying Procedures

Apples—Select mature, firm apples. Wash well. Pare, if desired, and core. Cut in rings or slices ⅛ to ¼ inch thick or cut in quarters or eighths. Soak in ascorbic acid, vinegar, or lemon juice for ten minutes. Remove from solution and drain well. Arrange in single layer on trays, pit side up. Dry until soft, pliable, and leathery; there should be no moist area in center when cut.

Apricots—Select firm, fully ripe fruit. Wash well. Cut in half and remove pit. Do not peel. Soak in ascorbic acid, vinegar, or lemon juice for ten minutes. Remove from solution and drain well. Arrange in single layer on trays, pit side up with cavity popped up to expose more flesh to the air. Dry until soft, pliable, and leathery; there should be no moist area in center when cut.

Bananas—Select firm, ripe fruit. Peel. Cut in 1/8-inch slices. Soak in ascorbic acid, vinegar, or lemon juice for ten minutes. Remove and drain well. Arrange in single layer on trays. Dry until tough and leathery.

Berries—Select firm, ripe fruit. Wash well. Leave whole or cut in half. Dip in boiling water thirty seconds to crack skins. Arrange on drying trays not more than two berries deep. Dry until hard and berries rattle when shaken on trays.

Cherries—Select fully ripe fruit. Wash well. Remove stems and pits. Dip whole cherries in boiling water thirty seconds to crack skins. Arrange in single layer on trays. Dry until tough, leathery, and slightly sticky.

Citrus peel—Select thick-skinned oranges with no signs of mold or decay and no color added to skin. Scrub oranges well with brush under cool running water. Thinly peel outer 1/16 to 1/8 inch of the peel; avoid white bitter part. Soak in ascorbic acid, vinegar, or lemon juice for ten minutes. Remove from solution and drain well. Arrange in single layers on trays. Dry at 130°F for one to two hours; then at 120°F until crisp.

Figs—Select fully ripe fruit. Wash or clean well with damp towel. Peel dark-skinned varieties if desired. Leave whole if small or partly dried on tree; cut large figs in halves or slices. If drying whole figs, crack skins by dipping in boiling water for thirty seconds. For cut figs, soak in ascorbic acid, vinegar, or lemon juice for ten minutes. Remove and drain well. Arrange in single layers on trays. Dry until leathery and pliable.

Grapes and black currants—Select seedless varieties. Wash, sort, and remove stems. Cut in half or leave whole. If drying whole, crack skins by dipping in boiling water for thirty seconds. If halved, dip in ascorbic acid or other antimicrobial solution for ten minutes. Remove and drain well. Dry until pliable and leathery with no moist center.

Melons—Select mature, firm fruits that are heavy for their size; cantaloupe dries better than watermelon. Scrub outer surface well with brush under cool running water. Remove outer skin, any fibrous tissue, and seeds. Cut into ¼- to ½-inch-thick slices. Soak in ascorbic acid, vinegar, or lemon juice for ten minutes. Remove and drain well. Arrange in single layer on trays. Dry until leathery and pliable with no pockets of moisture.

Nectarines and peaches—Select ripe, firm fruit. Wash and peel. Cut in half and remove pit. Cut in quarters or slices if desired. Soak in ascorbic acid, vinegar, or lemon juice for ten minutes. Remove and drain well. Arrange in single layer on trays, pit side up. Turn halves over when visible juice disappears. Dry until leathery and somewhat pliable.

Pears—Select ripe, firm fruit. Bartlett variety is recommended. Wash fruit well. Pare, if desired. Cut in half lengthwise and core. Cut in quarters, eighths, or slices ⅛ to ¼ inch thick. Soak in ascorbic acid, vinegar, or lemon juice for ten minutes. Remove and drain. Arrange in single layer on trays, pit side up. Dry until springy and suede-like with no pockets of moisture.

Plums and prunes—Wash well. Leave whole if small; cut large fruit into halves (pit removed) or slices. If left whole, crack skins in boiling water one to two minutes. If cut in half, dip in ascorbic acid or other antimicrobial solution for ten minutes. Remove and drain. Arrange in single layer on trays, pit side up, cavity popped out. Dry until pliable and leathery; in whole prunes, pit should not slip when squeezed.

Fruit Leathers

Fruit leathers are a tasty and nutritious alternative to store-bought candies that are full of artificial sweeteners and preservatives. Blend the leftover fruit pulp from making jelly or use fresh, frozen, or drained canned fruit. Ripe or slightly overripe fruit works best.

Chances are the fruit leather will get eaten before it makes it into the cupboard, but it can keep up to one month at room temperature. For storage up to one year, place tightly wrapped rolls in the freezer.

Ingredients

- 2 cups fruit
- 2 tsps lemon juice or ⅛ tsp ascorbic acid (optional)
- ¼ to ½ cup sugar, corn syrup, or honey (optional)

Directions

1. Wash fresh fruit or berries in cool water. Remove peel, seeds, and stem.
2. Cut fruit into chunks. Use 2 cups of fruit for each 13 x 15-inch fruit leather. Purée fruit until smooth.
3. Add 2 teaspoons of lemon juice or 1/8 teaspoon ascorbic acid (375 mg) for each 2 cups light-colored fruit to prevent darkening.

SPICES, FLAVORS, AND GARNISHES

To add interest to your fruit leathers, include spices, flavorings, or garnishes.

- **Spices to try**—Allspice, cinnamon, cloves, coriander, ginger, mace, mint, nutmeg, or pumpkin pie spice. Use sparingly; start with 1/8 teaspoon for each 2 cups of purée.
- **Flavorings to try**—Almond extract, lemon juice, lemon peel, lime juice, lime peel, orange extract, orange juice, orange peel, or vanilla extract. Use sparingly; try ⅛ to ¼ teaspoon for each 2 cups of purée.
- **Delicious additions to try**—Shredded coconut, chopped dates, other dried chopped fruits, granola, miniature marshmallows, chopped nuts, chopped raisins, poppy seeds, sesame seeds, or sunflower seeds.
- **Fillings to try**—Melted chocolate, softened cream cheese, cheese spreads, jam, preserves, marmalade, marshmallow cream, or peanut butter. Spread one or more of these on the leather after it is dried and then roll. Store in refrigerator.

4. Optional: To sweeten, add corn syrup, honey, or sugar. Corn syrup or honey is best for longer storage because these sweeteners prevent crystals. Sugar is fine for immediate use or short storage. Use ¼ to ½ cup sugar, corn syrup, or honey for each 2 cups of fruit. Avoid aspartame sweeteners as they may lose sweetness during drying.
5. Pour the leather. Fruit leathers can be poured into a single large sheet (13 x 15 inches) or into several smaller sizes. Spread purée evenly, about 1/8 inch thick, onto drying tray. Avoid pouring purée too close to the edge of the cookie sheet.
6. Dry the leather. Dry fruit leathers at 140°F. Leather dries from the outside edge toward the center. Larger fruit leathers take longer to dry. Approximate drying times are 6 to 8 hours in a dehydrator, up to 18 hours in an oven, and 1 to 2 days in the sun. Test for dryness by touching center of leather; no indentation should be evident. While warm, peel from plastic and roll, allow to cool, and rewrap the roll in plastic. Cookie cutters can be used to cut out shapes that children will enjoy. Roll, and wrap in plastic.

Vegetable Leathers

Pumpkin, mixed vegetables, and tomatoes make great leathers. Just purée cooked vegetables, strain, spread on a tray lined with plastic wrap, and dry. Spices can be added for flavoring.

Mixed Vegetable Leather

- 2 cups cored, cut-up tomatoes
- 1 small onion, chopped

HINTS

- Applesauce can be dried alone or added to any fresh fruit purée as an extender. It decreases tartness and makes the leather smoother and more pliable.
- To dry fruit in the oven, a 13 x 15-inch cookie pan with edges works well. Line pan with plastic wrap, being careful to smooth out wrinkles. Do not use waxed paper or aluminum foil.

- ¼ cup chopped celery
- Salt to taste

Combine all ingredients in a covered saucepan and cook over low heat 15 to 20 minutes. Purée or force through a sieve or colander. Return to saucepan and cook until thickened. Spread on a cookie sheet or tray lined with plastic wrap. Dry at 140°F.

Pumpkin Leather

- 2 cups canned pumpkin or 2 cups fresh pumpkin, cooked and puréed
- ½ cup honey
- ¼ tsp cinnamon
- ⅛ tsp nutmeg
- ⅛ tsp powdered cloves
- Blend ingredients well. Spread on tray or cookie sheet lined with plastic wrap. Dry at 140°F.

Tomato Leather

Core ripe tomatoes and cut into quarters. Cook over low heat in a covered saucepan, fifteen to twenty minutes. Purée or force through a sieve or colander and pour into electric fry pan or shallow pan. Add salt to taste and cook over low heat until thickened. Spread on a cookie sheet or tray lined with plastic wrap. Dry at 140°F.

How to Make a Woodstove Food Dehydrator

1. Collect pliable wire mesh or screens (available at hardware stores) and use wire cutters to trim to squares 12 to 16 inches on each side. The trays should be of the same size and shape. Bend up the edges of each square to create a half-inch lip.
2. Attach one S hook from the hardware store or a large paperclip to each side of each square (four clips per tray) to attach the trays together.
3. Cut four equal lengths of chain or twine that will reach from the ceiling to the level of the top tray. Use a wire or metal loop to attach the four pieces together at the top and secure to a hook in the ceiling above the woodstove. Attach the chain or twine to the hooks on the top tray.
4. To use, fill trays with food to dry, starting with the top tray. Link trays together using the S hooks or strong paperclips. When the foods are dried, remove the entire stack and disassemble. Remove the dried food and store.

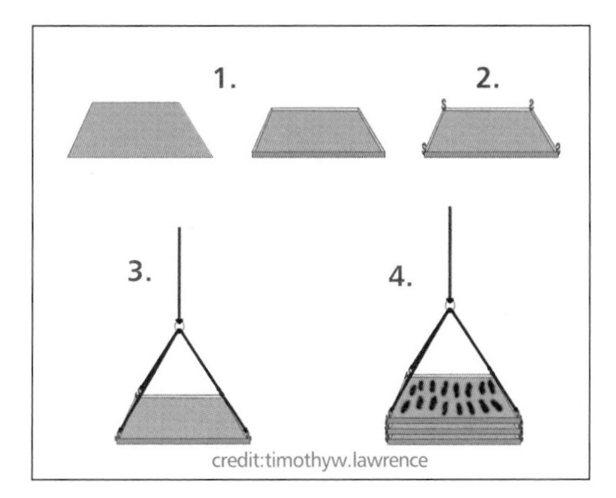

VINE DRYING

One method of drying outdoors is vine drying. To dry beans (navy, kidney, butter, great northern, lima, lentils, and soybeans), leave bean pods on the vine in the garden until the beans inside rattle. When the vines and pods are dry and shriveled, pick the beans and shell them. No pretreatment is necessary. If beans are still moist, the drying process is not complete and the beans will mold if not more thoroughly dried. If needed, drying can be completed in the sun, an oven, or a dehydrator.

Herbs

Drying is the easiest method of preserving herbs. Simply expose the leaves, flowers, or seeds to warm, dry air. Leave the herbs in a well-ventilated area until the moisture evaporates. Sun drying is not recommended because the herbs can lose flavor and color.

The best time to harvest most herbs for drying is just before the flowers first open when they are in the bursting bud stage. Gather the herbs in the early morning after the dew has evaporated to minimize wilting. Avoid bruising the leaves. They should not lie in the sun or unattended after harvesting. Rinse herbs in cool water and gently shake to remove excess moisture. Discard all bruised, soiled, or imperfect leaves and stems.

Dehydrator drying is another fast and easy way to dry high quality herbs because temperature and air circulation can be controlled. Preheat dehydrator with the thermostat set to 95 to 115°F. In areas with higher humidity, temperatures as high as 125°F may be needed. After rinsing under cool, running water and shaking to remove excess moisture, place the herbs in a single layer on dehydrator trays. Drying times may vary from one to four hours. Check periodically. Herbs are dry when they crumble, and stems break when bent. Check your dehydrator instruction booklet for specific details.

Less tender herbs—The more sturdy herbs, such as rosemary, sage, thyme, summer savory, and parsley, are the easiest to dry without a dehydrator. Tie them into small bundles and hang them to air dry. Air drying outdoors is often possible; however, better color and flavor retention usually results from drying indoors.

Tender-leaf herbs—Basil, oregano, tarragon, lemon balm, and the mints have a high moisture content and will mold if not dried quickly. Try hanging the tender-leaf herbs or those with seeds inside paper bags to dry. Tear or punch holes in the sides of the bag. Suspend a small bunch (large amounts will mold) of herbs in a bag and close the top with a rubber band. Place where air currents will circulate through the bag. Any leaves and seeds that fall off will be caught in the bottom of the bag.

Another method, especially nice for mint, sage, or bay leaf, is to dry the leaves separately. In areas of high humidity, it will work better than air drying whole stems. Remove the best leaves from the stems. Lay the leaves on a paper towel, without allowing leaves to touch. Cover with another towel and layer of leaves. Five layers may be dried at one time using this method. Dry in a very cool oven. The oven light of an electric range or the pilot light of a gas range furnishes enough heat for overnight drying. Leaves dry flat and retain a good color.

Microwave ovens are a fast way to dry herbs when only small quantities are to be prepared. Follow the directions that come with your microwave oven.

When the leaves are crispy, dry, and crumble easily between the fingers, they are ready to be packaged and stored. Dried leaves may be left whole and crumbled as used, or coarsely crumbled before storage. Husks can be removed from seeds by rubbing the seeds between the hands and blowing away the chaff. Place herbs in airtight containers and store in a cool, dry, dark area to protect color and fragrance.

Dried herbs are usually three to four times stronger than the fresh herbs. To substitute dried herbs in a recipe that calls for fresh herbs, use ¼ to ⅓ of the amount listed in the recipe.

Jerky

Jerky is great for hiking or camping because it supplies protein in a very lightweight form—plus it can be very tasty. A pound of meat or poultry weighs about four ounces after being made into jerky. In addition, because most of the moisture is removed, it can be stored for one to two months without refrigeration.

Jerky has been around since the ancient Egyptians began drying animal meat that was too big to eat all at once. Native Americans mixed ground dried meat with dried fruit or

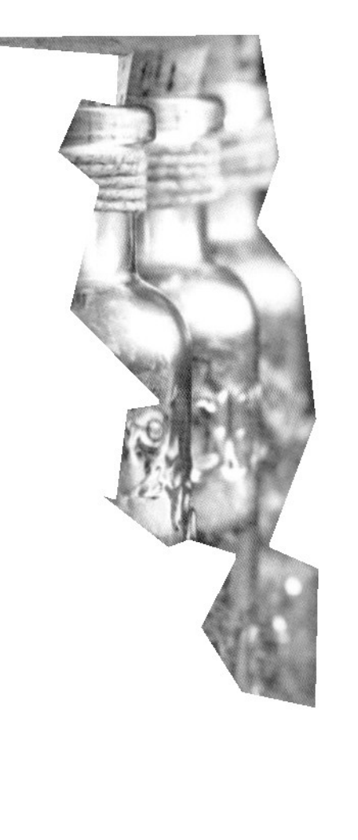

suet to make pemmican. *Biltong* is dried meat or game used in many African countries. The English word *jerky* came from the Spanish word *charque*, which means, "dried salted meat."

Drying is the world's oldest and most common method of food preservation. Enzymes require moisture in order to react with food. By removing the moisture, you prevent this biological action.

Jerky can be made from ground meat, which is often less expensive than strips of meat and allows you to combine different kinds of meat if desired. You can also make it into any shape you want! As with strips of meat, an internal temperature of 160°F is necessary to eliminate disease-causing bacteria such as *E. coli*, if present.

Food Safety

The USDA Meat and Poultry Hotline's current recommendation for making jerky safely is to heat meat to 160°F and poultry to 165°F before the dehydrating process. This ensures that any bacteria present are destroyed by heat. If your food dehydrator doesn't heat up to 160°F, it's important to cook meat slightly in the oven or by steaming before drying. After heating, maintain a constant dehydrator temperature of 130 to 140°F during the drying process.

According to the USDA, you should always:

- Wash hands thoroughly with soap and water before and after working with meat products.
- Use clean equipment and utensils.
- Keep meat and poultry refrigerated at 40°F or slightly below; use or freeze ground beef and poultry within two days, and whole red meats within three to five days.
- Defrost frozen meat in the refrigerator, not on the kitchen counter.
- Marinate meat in the refrigerator. Don't save marinade to reuse. Marinades are used to tenderize and flavor the jerky before dehydrating it.
- If your food dehydrator doesn't heat up to 160°F (or 165°F for poultry), steam or roast meat before dehydrating it.
- Dry meats in a food dehydrator that has an adjustable temperature dial and will maintain a temperature of at least 130 to 140°F throughout the drying process.

Preparing the Meat

1. Partially freeze meat to make slicing easier. Slice meat across the grain ⅛ to ¼ inch thick. Trim and discard all fat, gristle, and membranes or connective tissue.
2. Marinate the meat in a combination of oil, salt, spices, vinegar, lemon juice, teriyaki, soy sauce, beer, or wine.

Marinated Jerky

- ¼ cup soy sauce
- 1 tbsp Worcestershire sauce
- 1 tsp brown sugar
- ¼ tsp black pepper
- ½ tsp fresh ginger, finely grated
- 1 tsp salt
- 1½ to 2 lbs. of lean meat strips (beef, pork, or venison)

1. Combine all ingredients except the strips, and blend. Add meat, stir, cover, and refrigerate at least one hour.
2. If your food dehydrator doesn't heat up to 160°F, bring strips and marinade to a boil and cook for 5 minutes.
3. Drain meat in a colander and absorb extra moisture with clean, absorbent paper towels. Arrange strips in a single layer on dehydrator trays, or on cake racks placed on baking sheets for oven drying.

4. Place the racks in a dehydrator or oven preheated to 140°F, or 160°F if the meat wasn't precooked. Dry until a test piece cracks but does not break when it is bent (10 to 24 hours for samples not heated in marinade, 3 to 6 hours for preheated meat). Use paper towel to pat off any excess oil from strips, and pack in sealed jars, plastic bags, or plastic containers.

Freezing Foods

Many foods preserve well in the freezer and can make preparing meals easy when you are short on time. If you make a big pot of soup, serve it for dinner, put a small container in the refrigerator for lunch the next day, and then stick the rest in the freezer. A few weeks later, you'll be ready to eat it again and it will only take a few minutes to thaw out and serve. Many fruits also freeze well and are perfect for use in smoothies and desserts, or served with yogurt for breakfast or dessert. Vegetables frozen shortly after harvesting keep many of the nutrients found in fresh vegetables and will taste delicious when cooked.

Containers for Freezing

The best packaging materials for freezing include rigid containers such as jars, bottles, or Tupperware, and freezer bags or aluminum foil. Sturdy containers with rigid sides are especially good for liquids such as soup or juice because they make the frozen contents much easier to get out. They are also generally reusable and make it easier to stack foods in the refrigerator. When using rigid containers, be sure to leave headspace so that the container won't explode when the contents expand with freezing. Covers for rigid containers should fit tightly. If they do not, reinforce the seal with freezer tape. Freezer tape is specially designed to stick at freezing temperatures. Freezer bags or aluminum foil are good for meats, breads and baked goods, or fruits and vegetables that don't contain much liquid. Be sure to remove as much air as possible from bags before closing.

Effect of Freezing on Spices and Seasonings

- Pepper, cloves, garlic, green pepper, imitation vanilla, and some herbs tend to get strong and bitter.
- Onion and paprika change flavor during freezing.

HEADSPACE TO ALLOW BETWEEN PACKED FOOD AND CLOSURE

Headspace is the amount of empty air left between the food and the lid. Headspace is necessary because foods expand when frozen.

Type of Pack	Container with Wide Opening		Container with Narrow Opening	
	Pint	Quart	Pint	Quart
Liquid pack*	½ inch	1 inch	¾ inch	1½ inch
Dry pack**	½ inch	½ inch	½ inch	½ inch
Juices	½ inch	1 inch	1½ inch	1½ inch
*Fruit packed in juice, sugar syrup, or water; crushed or puréed fruit.				
**Fruit or vegetable packed without added sugar or liquid.				

FOODS THAT DO NOT FREEZE WELL

Food	Usual Use	Condition After Thawing
Cabbage*, celery, cress, cucumbers*, endive, lettuce, parsley, radishes	As raw salad	Limp, waterlogged; quickly develops oxidized color, aroma, and flavor
Irish potatoes, baked or boiled	In soups, salads, sauces, or with butter	Soft, crumbly, waterlogged, mealy
Cooked macaroni, spaghetti, or rice	When frozen alone for later use	Mushy, tastes warmed over
Egg whites, cooked	In salads, creamed foods, sandwiches, sauces, gravy, or desserts	Soft, tough, rubbery, spongy
Meringue	In desserts	Soft, tough, rubbery, spongy
Icings made from egg whites	Cakes, cookies	Frothy, weeps
Cream or custard fillings	Pies, baked goods	Separates, watery, lumpy
Milk sauces	For casseroles or gravies	May curdle or separate
Sour cream	As topping, in salads	Separates, watery
Cheese or crumb toppings	On casseroles	Soggy
Mayonnaise or salad dressing	On sandwiches (not in salads)	Separates
Gelatin	In salads or desserts	Weeps
Fruit jelly	Sandwiches	May soak bread
Fried foods	All except French-fried potatoes and onion rings	Lose crispness, become soggy

*Cucumbers and cabbage can be frozen as marinated products such as "freezer slaw" or "freezer pickles." These do not have the same texture as regular slaw or pickles.

- Celery seasonings become stronger.
- Curry develops a musty off-flavor.
- Salt loses flavor and has the tendency to increase rancidity of any item containing fat.
- When using seasonings and spices, season lightly before freezing, and add additional seasonings when reheating or serving.

How to Freeze Vegetables

Because many vegetables contain enzymes that will cause them to lose color when frozen, you may want to blanche your vegetables before putting them in the freezer. To do this, first wash the vegetables thoroughly, peel if desired, and chop them into bite-size pieces. Then pour them into boiling water for a couple of minutes (or cook longer for very dense vegetables, such as beets), drain, and immediately dunk the vegetables in ice water to stop them from cooking further. Use a paper towel or cloth to absorb excess water from the vegetables, and then pack in resealable airtight bags or plastic containers.

How to Freeze Fruits

Many fruits freeze easily and are perfect for use in baking, smoothies, or sauces. Wash, peel, and core fruit before freezing. To easily peel peaches, nectarines, or apricots, dip them in boiling water for fifteen to twenty seconds to loosen the skins. Then chill and remove the skins and stones.

Blanching Times for Vegetables	
Artichokes	3–6 minutes
Asparagus	2–3 minutes
Beans	2–3 minutes
Beets	30–40 minutes
Broccoli	3 minutes
Brussels sprouts	4–5 minutes
Cabbage	3–4 minutes
Carrots	2–5 minutes
Cauliflower	6 minutes
Celery	3 minutes
Corn (off the cob)	2–3 minutes
Eggplant	4 minutes
Okra	3–4 minutes
Peas	1–2 minutes
Peppers	2–3 minutes
Squash	2–3 minutes
Turnips or Parsnips	2 minutes

Berries should be frozen immediately after harvesting and can be frozen in a single layer on a paper towel-lined tray or cookie sheet to keep them from clumping together. Allow them to freeze until hard (about three hours) and then pour them into a resealable plastic bag for long-term storage.

Some fruits have a tendency to turn brown when frozen. To prevent this, you can add ascorbic acid (crush a vitamin C in a little water), citrus juice, plain sugar, or a sweet syrup (one part sugar and two parts water) to the fruit before freezing. Apples, pears, and bananas are best frozen with ascorbic acid or citrus juice, while berries, peaches, nectarines, apricots, pineapple, melons, and berries are better frozen with a sugary syrup.

How to Freeze Meat

Be sure your meat is fresh before freezing. Trim off excess fats and remove bones, if desired. Separate the meat into portions that will be easy to use when preparing meals and wrap in foil or place in resealable plastic bags or plastic containers. Refer to the chart below to determine how long your meat will last at best quality in your freezer.

Meat	Months
Bacon and sausage	1 to 2
Ham, hotdogs, and lunchmeats	1 to 2
Meat, uncooked roasts	4 to 12
Meat, uncooked steaks or chops	4 to 12
Meat, uncooked ground	3 to 4
Meat, cooked	2 to 3
Poultry, uncooked whole	12
Poultry, uncooked parts	9
Poultry, uncooked giblets	3 to 4
Poultry, cooked	4
Wild game, uncooked	8 to 12

Part 3

Country Crafts

Spring	166
Summer	175
Autumn	184
Winter	196

Happiness is not in the mere possession of money; it lies in the joy of achievement, in the thrill of the creative effort.

—President Franklin Delano Roosevelt

Spring

Springtime lends itself to creativity. As new flowers push through the soil and birds begin to gather materials for their nests, you may find yourself eager to start new projects, too. It is a busy season; days are filled with preparing gardens and airing out the house, weekends are packed with weddings and graduations. But a rainy afternoon or a quiet Sunday may give you the opportunity for crafting you crave. Here is a smattering of ideas to give your inspiration some direction, whether you're creating a gift, an accent for your home, or a keepsake for a special celebration.

Springtime Wreath

Grapevines make an attractive and natural base for this welcoming wreath. Begin shaping the vines soon after cutting. If you do need to store them for more than a day before using them, remove the leaves and soak the vines for several hours before beginning the wreath to make them more pliable.

- Grapevines
- Decoration such as moss, baby's-breath, etc.
- Florist tape

1. Cut several lengths of vine, keeping them as long as you can manage. Remove the leaves, but don't trim the tendrils.
2. Start with the thicker end of the vine and form a circle. Then begin coiling a second circle, winding and twisting

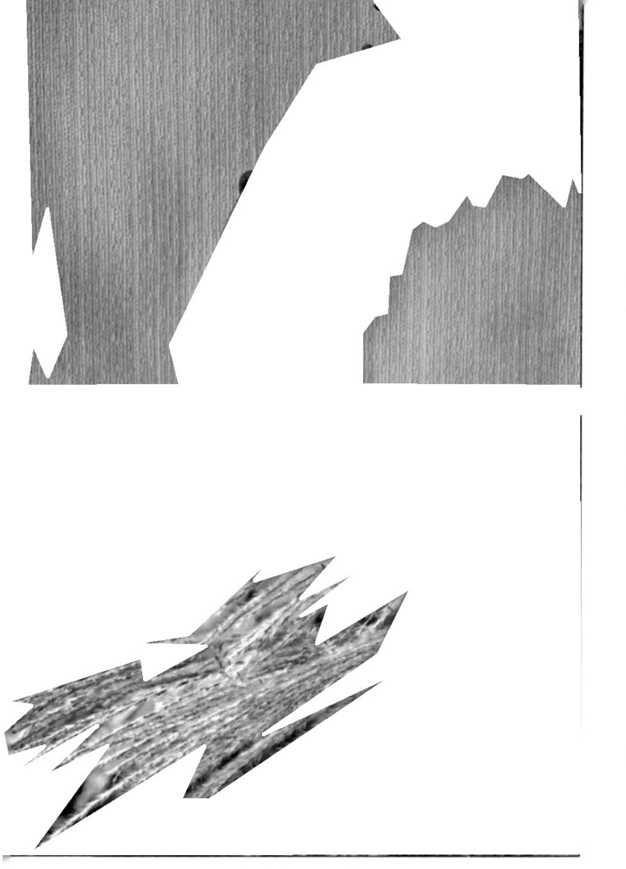

To dry flowers or grasses for use in wreaths or other decorations, cut them slightly before they reach their peak, leaving long stems. Tie the stems together and hang them upside down for several weeks in a dry, dark area.

it around the first. Continue until you are almost at the end of the vine and then wind the end around and around, using the vine tendrils or florist tape to secure the end. Add additional vines, winding them around the first circle and tucking the ends into the center.

3. Add moss, baby's-breath, yarrow, ferns, feathers, decorative grasses, ribbon, etc. Tuck the ends into the vines and use florist tape as needed for added security.

Blown Eggs

If you want to keep your egg creations to display year after year, blow out the eggs before decorating them. Blown eggs are more fragile than hardboiled eggs, but they won't ever spoil.

- Eggs
- Needle
- Toothpick
- Tiny straw or a syringe

1. Use a needle to poke one hole on each end of the egg. Insert a toothpick and wiggle it around to help widen the holes slightly. Stick the needle back into one hole and move it around inside the egg until the yolk breaks.

PLANTS THAT DRY WELL FOR DECORATIVE USE

Flowers	Grasses	Herbs
Acrolinium, Baby's-breath, Bachelor's button, Bells of Ireland, Cockscomb, Coneflower, Delphinium, Foxglove, Globe amaranth, Goldenrod, Heather, Hydrangea, Larkspur, Statice, Strawflower, Yarrow	Bristly foxtail, Cattails, Eulalia grass, Fountain grass, Hare's-tail grass, Indian grass, Northern sea oats, Pampas grass, Plume grass, Quaking grass, Spike grass, Squirrel-tail grass, Switch grass, Wheatgrass	Chamomile, Chives, Dill, Eucalyptus, Fennel, Lavender, Lemongrass, Rosemary, Sage, St. John's wort, Thyme

2. Insert the straw into one hole and blow through it until the insides of the egg drain out, or use a syringe to draw out the egg. Rinse thoroughly to wash away any remaining egg residue.
3. Bake the eggshell at 400°F for ten minutes.

Leaf- or Flower-Stenciled Eggs

These eggs make a stunning centerpiece when displayed on a plate or in a basket. If you intend to eat the eggs later, use only food coloring to dye them. However, if you are using blown eggs or you do not intend to eat the eggs, fabric dye, or other natural dyes, and stencil paint will produce richer, more vibrant colors.

- Food coloring, fabric dye, or other natural dye (prepared accordingly)
- Large white eggs
- White glue
- Leaves, ferns, or small flowers
- Small paintbrush
- Wide, stiff paintbrush
- Stencil paint (if you do not intend to eat the egg later)

1. Fill a deep bowl or wide glass half full with the prepared dye solution.
2. Immerse an egg in the dye mixture and allow it to sit for a few minutes, or until the egg has reached the desired color. Rinse the egg in cold water and allow it to dry.
3. Using a small paintbrush, paint a thin layer of glue on the back of your leaf or fern. Stick the leaf or fern to the egg.
4. Dip a wide paintbrush in undiluted food coloring (if you intend to eat the eggs later) or stencil paint. Blot the brush on newspaper or paper towel to get rid of excess paint. Dab the ends of the bristles up and down over the leaf, allowing the first layer of color to show through to create a dappled effect.
5. When the egg is completely dry, peel the leaf away from the egg.

Natural Dyes for Easter Eggs

There are many ingredients from nature you can use to dye your Easter eggs. The colors may be more subdued than if you use food coloring or paints, but you can achieve some beautiful pastels with berries, flowers, and other plants and foods. Mix dyestuff or the finished dyes to make more color variations. Note that liquid dyestuffs, such as grape juice, do not need to be simmered with water as described below. Simply add the vinegar and use!

- Dyestuff (see chart below)
- Water
- White vinegar
- Cooking oil or mineral oil

1. Place a handful or two of the dyestuff of your choice (or a couple of tablespoons if using herbs or spices) in a saucepan.

2. Add water until the dyestuff is fully submerged. Simmer on low for about fifteen minutes, or until the desired color is reached. Keep in mind that the eggs will turn out paler than the dye appears in the pan.
3. Strain dye into a liquid measuring cup. Add 2 tablespoons of white vinegar for every cup of dye. The dye is now ready for use.
4. After the dyed eggs are dry, rub the eggs with cooking oil or mineral oil to give them a glossier sheen.

Terrariums

Terrariums are miniature ecosystems that you can create and keep indoors. They're a wonderful way to learn about gardening on a small scale and can add interest to your home décor and oxygen to the air you breathe. Terrariums can contain only plants or can be homes for lizards, turtles, or other small animals. If you do wish to make your terrarium a home for a pet, be sure to include the proper shelter for the animal and a way

to provide food and water. Some terrariums even include waterfalls so that animals can have a constant supply of fresh water! The size of the terrarium can be as large as a fish tank or as small as a thimble. Bowls, teapots, jars, and bottles have all been successfully transformed into miniature indoor gardens. Terrariums can be fully enclosed or can have an open top to allow fresh air to circulate. Because one of the benefits of a terrarium is the oxygen that the plants contribute to the air, these directions are for an open top terrarium.

- Container (preferably a clear glass container, so you can easily see your miniature garden)

Color	Items to Dye With
Blue	Blueberries, red cabbage, purple grape juice
Brown	Chili powder
Brown or Beige	Coffee grounds, black walnut shells, black tea leaves
Brown Gold	Dill seeds
Green	Spinach leaves, liquid chlorophyll
Gray	Purple or red grape juice, beet juice
Lavender	Purple grape juice, violet blossoms plus a little lemon juice, Red Zinger tea
Orange	Yellow onion skins, carrots, paprika
Pink	Beets, crushed cranberries or cranberry juice, crushed raspberries, grape juice
Red	Lots of red onion skins, pomegranate juice, canned cherries, crushed raspberries
Violet or Purple	Violet blossoms, hibiscus tea, small quantity of red onion skins, red wine
Yellow	Orange or lemon peels, carrot tops, chamomile tea, celery seed, green tea, ground cumin, ground turmeric, saffron

- Coarse sand or pebbles
- Sphagnum moss
- Soil
- Seeds or seedlings
- Water
- Ornaments (optional)

1. Place a ½-inch to 1-inch layer of coarse sand or pebbles in the bottom of your container. This will help the soil to drain properly.
2. Add a layer of moss over the pebbles. The moss acts as a filter, allowing the soil to drain but not to seep down into the pebbles.
3. Pour the soil over the moss and spread evenly. How much soil you use will depend on how big your container is and how large the plants will grow. Pat the dirt down firmly.
4. Plant the seeds or seedlings. Think carefully about how you want the plants to be arranged. You may want taller plants in the center and shorter ones toward the outside so that you can see them all. Add pretty stones, pinecones, figurines, or other ornaments, if desired.
5. Place your terrarium in a sunny spot and water it regularly.

Mosaic Flowerpots

Spring is the time to start planting seeds so you'll have seedlings to transplant to the garden come summer. Make your pots unique by decorating them yourself with bits of beach glass, pottery, sea shells, or beads. This is a great project to do with kids, but be careful of sharp pieces of glass or pottery.

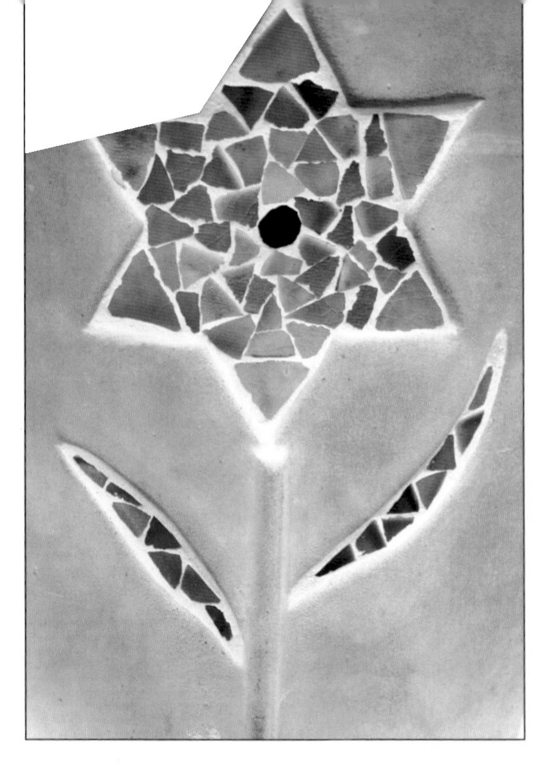

TIP

Look for nontoxic grout online or at your local hardware store. Alternatively, you can make your own grout by mixing one part Portland cement to two parts sand in a tub. Add water slowly while stirring until the mixture is thick like mud. If desired, add natural iron oxide pigments to the grout for a more colorful background for your mosaic.

For an even simpler "faux grout," mix two parts white sand to one part white glue. Add acrylic paint or concentrated natural dyes as desired. The "faux grout" won't be as strong or as smooth as real grout, but it will work in a pinch.

- Putty knife
- Ceramic tile grout
- Terra-cotta flowerpot
- Pieces of beach glass, broken pottery or mirror, tile, beads, charms, etc.

1. Use the putty knife to spread a thick layer of grout around the outside of the flowerpot (at least ¼-inch thick).
2. Press the pieces into the grout and add more grout around each piece to cover any sharp or rough edges.
3. Allow pot to dry thoroughly. Then wipe away any grout residue from the pieces with a damp sponge.

Homemade Weddings

Weddings are wonderful, memorable, and very personal events. Every couple is unique and weddings are an opportunity to celebrate their distinct experiences of love, beauty, and joy. Couples who do some or all of the wedding preparations on their own or with the help of friends and family (rather than relying entirely on professionals) often find their day especially meaningful—and they're less likely to start off their married life in debt.

There are endless ways to add homemade touches to a wedding. The invitations, flowers, centerpieces, favors, food, and even the attire can be made or arranged without the help of professional florists, caterers, and other service providers. Couples should think about what aspects of the wedding matter most to them, what they would most enjoy doing on their own, and how much time they can realistically dedicate to wedding preparations. Delegation is also key—friends and family are often more than happy to be a part of the celebration by assisting with certain tasks.

Bouquets

Once you've chosen the flowers you want in your bouquets (often based on color and season), think about what shape the bouquets will be and how you will arrange them. Smaller bouquets can be secured with a wide ribbon tied around the stems a little below the flower heads. For larger bouquets or for flowers with rough or thorny stems, you may want to bind the stems together with a ribbon that covers the length of the stems.

- Flowers
- Floral tape
- Ribbon
- Corsage pins

1. Arrange the flowers in a bunch, cutting the stems evenly at the bottom. Wrap a piece of floral tape around the stems just below the blossoms. Cut a piece of ribbon about five feet long (or longer for very long stems).
2. Begin wrapping the ribbon around the stems just below the blossoms, leaving a tail of ribbon about two feet long.

3. Wrap the ribbon around and around the stems, working your way down and making sure the ribbon lies flat. When you reach the bottom, wrap over the bottom of the stems and work your way back up to the top.
4. Now you should have two ribbon tails that you can tie into a full bow. Use pearl-headed corsage pins to secure the ribbon in place. Trim the tails if desired.

Hanging Flower Pomander

Flower pomanders should be made as late as possible before the celebration, since the flowers will not be getting adequate water to stay fresh for as long as a bouquet in a vase would. If making the pomanders the day before the wedding, keep them refrigerated as long as possible. Choose flowers with stiff stems such as roses or carnations, as they'll be easier to insert into the ball.

- Ball of floristry foam
- 2½-inch-wide ribbon

- Straight pins or a glue gun
- Scissors
- Flowers

1. Soak floral foam in water until saturated and then set aside on a towel and allow to dry until just damp.
2. Wrap the ribbon once around the outside of the foam ball, pinning it at every inch or securing it to the ball with hot glue. Leave a foot of ribbon loose on each end.
3. Trim your flowers, leaving about an inch of stem for smaller blossoms and two to three inches for larger blossoms. Push the stems all the way into the ball, following the edge of the ribbon. Then do a line of flowers around the circumference so that the ball is divided into corners. Finally, fill in the four sectors, adding ferns or other greenery as desired to close up any spaces. If desired, only cover the top two-thirds of the ball with flowers and line the bottom third with large green leaves, secured with pins.
4. Tie the two ends of the ribbon together in a tight knot and hang the pomander in a door or window or on a chandelier, or carry one down the aisle instead of a bouquet.

Wedding Cake

Your cake can be any flavor you like, and any shape, for that matter. But to make a traditional tiered wedding cake, follow the steps below. Be sure to do at least one test run well before the wedding.

- ½-inch-thick plywood or large cutting board
- Parchment paper
- Tape
- Frosting
- Cake tiers
- Corrugated cardboard
- Tinfoil
- Knife or toothpick

- ¼-inch-thick dowels
- Heavy duty shears
- Cake decorations

1. Prepare the cake board. Cover a ½-inch-thick piece of plywood or a large cutting board with parchment paper, taping it to the underside of the board. The board should be at least 4 inches larger than the largest tier of your cake. This board will stay with your cake until it's eaten, so do a neat job of covering it.
2. Frost and fill the bottom, largest cake tier and place it in the center of the cake board.
3. Cut out two circles of corrugated cardboard that are the same size as the second tier, and two circles the same size as the third tier. Tape the same-sized circles together so that they are double thickness. Cover each cardboard circle with tinfoil, making the foil as smooth as possible.
4. Gently place the medium-sized cardboard circle onto the center of the bottom tier and mark around the circle with a knife or toothpick. Now you can remove the cardboard and still see the circle.
5. Push a ¼-inch-thick dowel into the very center of the cake. Mark the height of the cake on the dowel and remove it. Use

A few fresh blossoms floating in a bowl make a simple and elegant centerpiece.

EDIBLE FLOWERS FOR CAKE DECORATING

Any nontoxic flowers can be used to decorate your cake, as long as you remove the flowers before serving. If you don't want to remove the flowers, choose edible flowers such as the following:

- Apple blossoms
- Orange blossoms
- Cornflowers
- Pansies
- Violets
- Nasturtiums
- Oregano
- Lavender
- Carnations
- Clover
- Calendula
- Jasmine
- Hollyhock
- Impatiens
- Lilacs
- Peonies
- Roses

Refer to page 179 for directions for making sachets. Sachets filled with candy, a few homemade cookies, or handmade soap make lovely favors.

heavy duty shears to cut eight dowels at that length. Insert the dowels into the cake at even intervals around the circle marked on the bottom tier.
6. Place the second cake tier on the medium-sized cardboard circle and fill and frost it. Place it carefully on the center of the bottom tier.
7. Repeat steps 5 and 6 with the top layer. To help keep the layers from sliding, sharpen one long dowel with a knife and insert it through the center of the top layer straight down through all the layers. Trim the dowel so that it's even with the top of the cake.
8. Decorate the cake with piping around the bottom of each tier, marzipan fruit, or fresh fruit or flowers.

Programs for the Ceremony

A handmade program will give your guests a unique keepsake by which to remember your ceremony. To make your own paper, refer to page 203.

- Paper in various colors
- Pen
- Paste
- Decorations

1. Cut a piece of sturdy, attractive paper to 7 x 11 inches, or desired size.
2. Write or print the text on paper of a contrasting color or texture, making sure it fits on a rectangle that is slightly smaller than the original cut rectangle (about 6 x 10 inches).
3. Cut out the second rectangle and paste it to the first, centering it carefully. Allow to dry and then fold the program in half.
4. If desired, paste ribbon, pressed flowers, or other accents to the outside of the program.

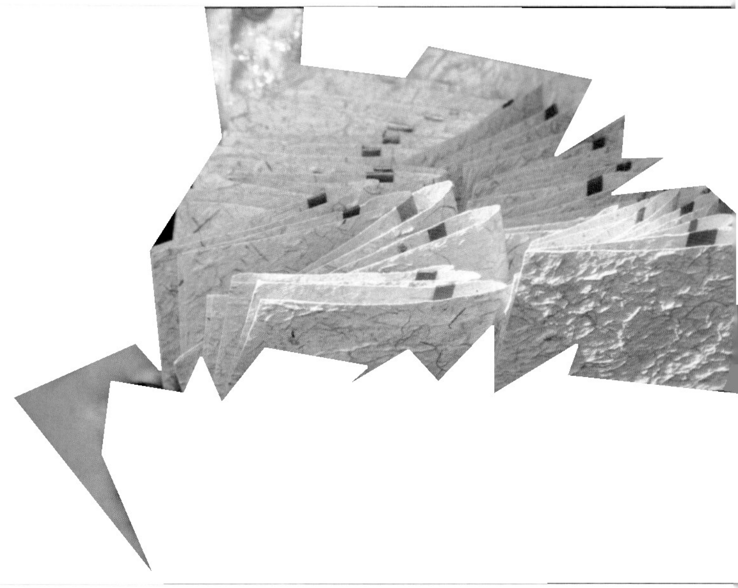

Slip flowers into folded napkins to add color to simple table settings.

Summer

Ripe berries, gardens full of flowers, warm breezes—summer is the season of simple pleasures. What better way to enjoy it than by relaxing in a handmade hammock, flying a kite you made yourself, or concocting a potpourri recipe that will remind you of the sweet smells of summer all year round?

Berry Ink and Quill Pens

During the Civil War, soldiers made ink out of berry juice and used feathers or corn stalks to write important letters. Use berry ink and a quill pen for special invitations, place cards, or just for fun. The vinegar helps the ink to retain its color and the salt acts as a natural preservative. Store extra ink in an airtight jar.

- Strainer
- Bowl
- Spoon
- ¼ cup berries (raspberries, strawberries, currants, or any other brightly colored berry will work)
- ¼ teaspoon vinegar
- ¼ teaspoon salt
- Bird feather
- Scissors
- Paper

1. Place the strainer over the bowl and use the back of a spoon to squish the berries so that the juice runs into the bowl. Once all the juice is extracted, discard the berry remains (or add them to your compost pile).
2. Add the vinegar and salt to the berry juice. Mix thoroughly.
3. Cut the sharp tip of the feather off at an angle. Dip the quill into the ink and begin writing. Practice on scrap paper before attempting an important project.

Potpourri from Your Garden

Rose petals and sweet geranium leaves are the primary ingredients in most potpourri. You can also add lavender, sweet verbena leaves, bay leaves, rosemary, dried orange peels, or pine needles. Orrisroot powder will help preserve the scent of the other ingredients.

The best time to collect flowers, seeds, or roots is in the morning, just after the dew has evaporated. Be careful not to bruise flower petals when gathering them, as damaged flowers will lose their scent.

- Flower petals, flower heads, stems, roots, or fruit (see chart below)
- Screen
- String
- Food dehydrator, oven, or solar oven
- Spices and orrisroot
- Essential oil

1. To dry individual petals or flower heads, lay them on a screen and leave them in a dry place, out of direct sunlight, for about two weeks.
2. For stems or roots, bunch them together, tie with string, and hang upside down for one to two weeks.
3. Flowers or fruit (such as citrus slices or peels) can also be dried in a food dehydrator set at its lowest temperature, or in the oven set to 180°F. Drying in the oven takes several hours. Leave the oven door open slightly to allow moisture to escape. Using a solar oven will take longer but is much more energy efficient!
4. To enhance the fragrance of your potpourri, add spices and orrisroot (available online or from many florists) to your final mixture. Gather violet powder, ground allspice, ground cloves, ground mixed spice, ground mace, whole mace, and/or whole cloves.

5. Once all the components are thoroughly dried, you can mix them together. To make the potpourri smell stronger, you can add a few drops of essential oil and store the mixture in a crock with a tight-fitting lid or a lidded jar for a few weeks or even months. As the mixture sits, the fragrance tends to become much richer.

The chart below shows the main types of fragrances and the ingredients associated with them. When experimenting with creating a recipe, first decide whether you want your potpourri to be primarily spicy, sweet, fruity/citrusy, or earthy. Use mostly ingredients from that category and then add smaller amounts from the other categories, if desired.

"You Choose" Potpourri Recipe

Potpourri can be comprised of a mix of flowers, leaves, and herbs—depending on which of these are available to you. The ingredients in this recipe are interchangeable and can all be used together or you can pick and choose which you want to use in making your own potpourri.

Essential Ingredients

- 1 ounce orrisroot
- 1 ounce allspice
- 1 ounce bay salt
- 1 ounce cloves

Assorted Ingredients (to add as you like)

- Rose petals
- Lavender

POTPOURRI CHART

Fragrance Category	Ingredients
Spicy	Allspice Berry, Bay, Cardamom, Caraway, Cinnamon, Clary Sage, Clove, Coriander, Cumin, Fir, Frankincense, Hyssop, Myrrh, Neroli, Nutmeg, Rosewood
Sweet	Anise, Clary Sage, Coriander, Frankincense, Geranium, Jasmine, Lavender, Rose, Sandalwood, Vanilla
Fruity/ Citrusy	Bergamot, Berries, Lemon, Lemongrass, Orange, Ylang ylang, Wintergreen,
Earthy	Cedarwood, Chamomile, Eucalyptus, Fir, Juniper, Myrrh, Patchouli, Pine, Rosemary, Spruce

- Lemon plant
- Verbena
- Myrtle
- Rosemary
- Bay leaf
- Violets
- Thyme
- Mint
- Essence of lemon
- Essence of lavender

Mix the orrisroot, allspice, bay salt, and cloves together. Combine this with about twelve handfuls of the dried petals and leaves and store in an airtight jar or bowl. A small quantity of essence of lemon and/or lavender may be added but these are not necessary. Let the mixture stand for a few weeks. If it becomes too moist, add additional powdered orrisroot. Once the potpourri is dry and very fragrant, parcel it into bowls to set around the house or give as gifts in jars or sachets.

Rose Potpourri

Ingredients

- 1 lb rose petals (already pressed and from a jar)
- Dried lavender (any proportion you like)
- Lemon verbena leaves (any proportion you like)
- A dash of orange thyme
- A dash of bergamot
- 1 dozen young bay leaves (dried and broken up)
- A pinch of musk
- 2 ounces orrisroot, crushed
- 1 ounce cloves, crushed
- 1 ounce allspice, crushed
- ½ ounce nutmeg, crushed
- ½ ounce cinnamon, crushed

Combine all the ingredients together in a large crock, jar, or bowl with a lid. Seal and allow it to sit for a few weeks, until the aroma is to your liking. Parcel into small bowls to fragrance rooms or give as gifts.

A Simple Recipe for Sachet Potpourri

If you are unable to procure large amounts of petals and leaves, here is a simple recipe, using oils as substitutes, that makes fine potpourri sachets.

Ingredients

- 2 drams alcohol
- 10 drops bergamot
- 20 drops eucalyptus oil
- 4 drops oil of roses
- ½ tsp cloves
- 1 ounce orrisroot
- ¼ tsp cinnamon
- ½ tsp mace
- 1 ounce rose sachet powder

Mix these ingredients together in a large stone crock or in a large glass bowl. When the ingredients are thoroughly mixed, store the potpourri in small wooden boxes or sachets and place them around the house. The potpourri gives off a pleasing fragrance to any room or drawer.

Sachet Bags

Sachets are small bags filled with sweet-smelling potpourri. Hang them in a closet or tuck them in a drawer to lend your clothes a gentle fragrance.

- Fabric
- Scissors
- Needle and thread or sewing machine
- Ribbon or string

1. Cut two 3 x 5-inch rectangles of fabric. Hem one of the 3-inch sides of both rectangles. Stack the pieces one on top of the other, placing the hemmed edges together and lining up all sides exactly. The right sides of the fabric should be facing toward each other.
2. Sew along three sides of the fabric, about ¼ inch from the edges. Leave one of the 3-inch sides unsewn so that the bag has an opening.
3. Turn the bag inside out so that the right side of the fabric is visible and the seams are hidden. Fill the bag half full with potpourri and tie a ribbon or string around the top.

Preserving Flowers

Pressed Flowers and Leaves

Pressed flowers can be used to decorate stationery, handmade boxes, bookmarks, scrapbooks, or picture frames. Kids can glue pressed wildflowers to a blank book and add species names and descriptions to make their own field guides.

- Large book or newspapers
- Blotting paper
- Weights
- Leaves, ferns, or flowers

1. Have a large book or a quantity of old newspapers and blotting paper, and several weights ready.
2. Use the newspapers for leaves and ferns. Blotting paper is best for the flowers. Both the flowers and leaves should be fresh and without moisture. Place them as nearly as possible in their natural positions in the book or papers, and press, allowing several thicknesses of paper between each layer.
3. Remove the flowers and leaves onto dry papers each day until they are perfectly dried.

Some flowers, like orchids, must be immersed—all but the flower head—in boiling water for a few minutes before pressing, to prevent them from turning black.

In order to preserve your flowers forever, get a blank book or pieces of stiff, white paper on which to mount your preserved flowers and leaves. You can glue them down to the paper with hot glue or regular Elmer's glue. The sooner you mount the specimens, the better. Place them carefully on the paper and, beneath each flower or leaf, you may want to write the name of the plant, where it was found, and the date.

Natural Wax Flowers

- Paraffin
- Saucepan
- Fresh flowers or leaves and ferns
- Wax paper
- Iron

1. To make wax flowers, dip the fresh buds and blossoms in paraffin that is just hot enough to be liquefied. First dip the stems of the flowers. When these have cooled and hardened, then dip the flowers or sprays. Be sure to hold them by the stalks and move them gently.
2. When they are completely covered, remove the flowers from the wax and shake them lightly in order to throw off the excess wax. Allow the flowers to dry completely. The flowers will keep their beautiful coloring and natural forms, and even their fragrance for a short while.
3. For leaves, ferns, or flat flowers, you can place the plant between two sheets of wax paper and run a hot iron over the paper. Allow the paper to cool slightly and then carefully remove it.

Pine Cone Bird Feeder

TIP

Make your own bird seed mix by combining millet, cracked corn, sunflower seeds, safflower seeds, and thistle seeds. Birds also love bits of stale bread, cereals, dried fruit, and nuts.

To attract wild birds to your yard, all you need is a pine cone, a bunch of seeds, and a little peanut butter. Keep in mind that wild animals also enjoy backyard treats, so you may end up attracting bears, raccoons, or other unwanted fauna. But, in the meantime, the birds will appreciate your generosity!

- Cord or string (at least 2 feet long)
- Large pine cone
- Peanut butter
- Birdseed

1. Loop the cord around the top petals of the pine cone and tie it tightly. Then

spoon a little peanut butter between each layer of pine cone petals.
2. Roll the pine cone in the birdseed (you can spread the seeds in a pie dish or on a sheet of waxed paper first.
3. Hang the feeder in a tree.

QUALITIES OF A GOOD KNOT

1. It can be tied quickly.
2. It will hold tightly.
3. It can be untied easily.

Tying Knots

Knowing how to tie a variety of knots is invaluable, especially if you are involved in boating, rock climbing, fishing, or other outdoor activities.

Strong knots are typically those that are neat in appearance and are not bulky. If a knot is tied properly, it will almost never loosen and will still be easy to untie when necessary.

The best way to learn how to tie knots effectively is to sit down and practice with a piece of cord or rope. Listed below are a few common knots that are useful to know:

- **Bowline knot:** Fasten one end of the line to some object. After the loop is made, hold it in position with your left hand and pass the end of the line up through the loop, behind and over the line above, and through the loop once again. Pull it tightly and the knot is now complete.

THREE PARTS OF A ROPE

1. **The standing part:** this is the long, unused part of the rope.
2. **The bight:** this is the loop formed whenever the rope is turned back.
3. **The end:** this is the part used in leading.

- **Clove hitch:** This knot is particularly useful if you need the length of the running end to be adjustable.

- **Halter:** If you need to create a halter to lead a horse or pony, try this knot.

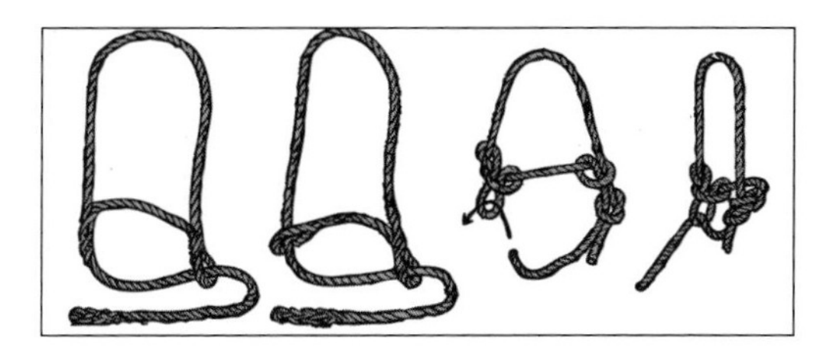

- **Sheepshank knot:** This is used for shortening ropes. Gather up the amount to be shortened and then make a half hitch around each of the bends.

- **Slip knot:** Slip knots are adjustable, you can tighten them around an object after they're tied.

- **Square/reef knot:** This is the most common knot for tying two ropes together.

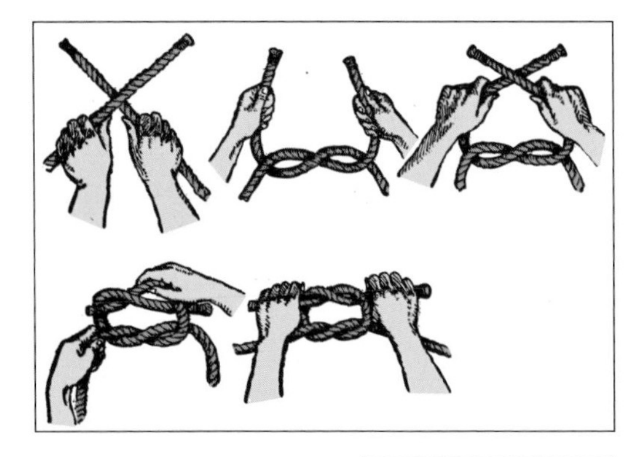

- **Timber hitch:** If you need to secure a rope to a tree, this is the knot to use. It is easy to untie, too.

- **Two half hitches:** Use this knot to secure a rope to a pole, boat mooring, washer, tire, or similar object.

Boomerangs

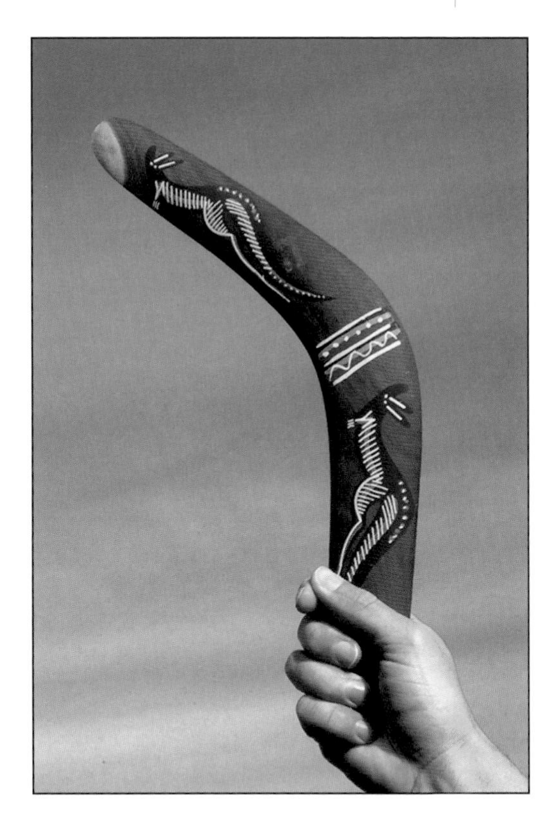

In order to make a boomerang, scald a piece of well-seasoned elm, ash, or hickory plank (free from any knots) in a pot of boiling water. Allow the wood to remain in the water until it becomes pliable enough to bend into a slight V-shaped form. When the wood has assumed the proper shape, nail on the side pieces to hold the wood in position until it is thoroughly dry. After the plank is completely dry, the side pieces can be removed—the wood will keep the curved shape.

Saw the wood into as many pieces as it will allow and each piece will become a boomerang. If the edges are very rough, trim them with a pocketknife and scrape them smooth. You can use a large file to help shape the boomerang. The efficiency of your boomerang (how well it soars and returns to you) will vary in each piece, depending on the curvature.

HOW TO THROW A BOOMERANG

Grasp the boomerang near one end and hold it like a club. Make sure the concave side is turned away from you and the convex side is toward you. Find something to take aim at and then throw the boomerang at the object. If the boomerang is well made, it should return to you after its flight. Be careful not to throw the boomerang when others are close by—it may end up hitting them and it can leave a bad welt. It is best to throw your boomerang in a large, open field by yourself.

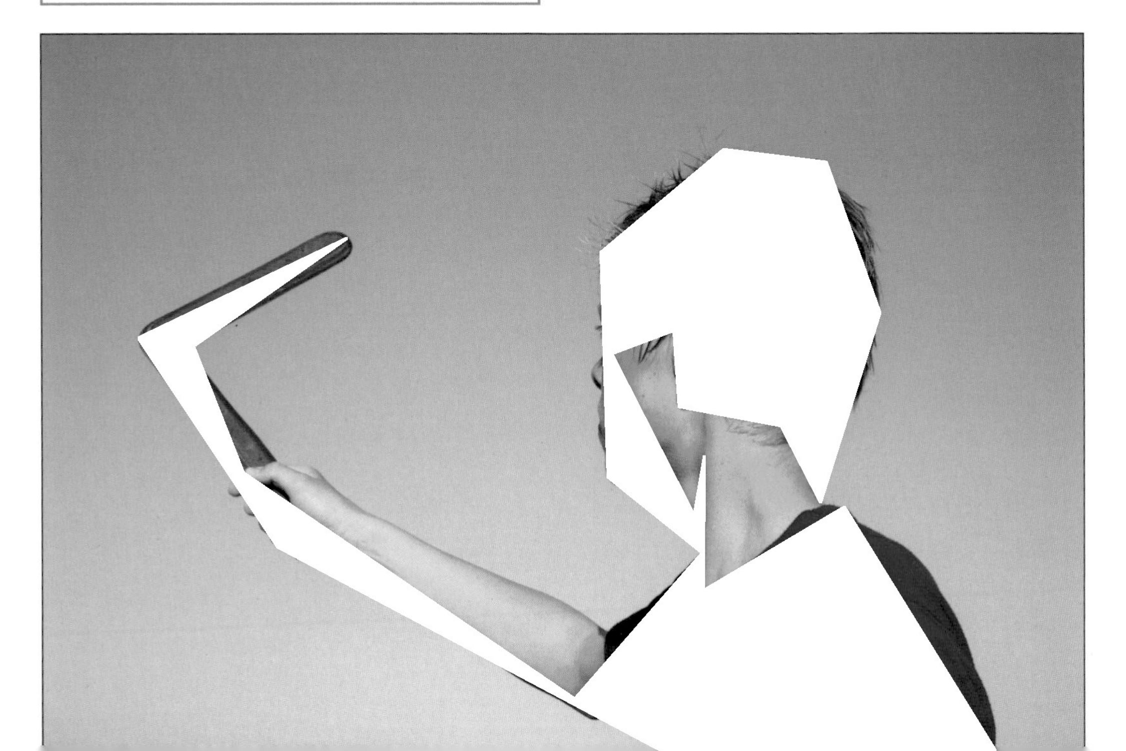

Autumn

Fall days may be filled with harvesting the gardens and savoring the fleeting rays of sunshine, but a cool autumn evening is perfect for cozy craft projects. Try dipping candles, making fragrant soaps, or weaving baskets from reeds or rushes. The kids will have fun making dolls from the husks leftover from late summer corn.

Handmade Candles

Making candles is a great activity for a fall afternoon. Simple beeswax candles can be completed in a few minutes, but give yourself several hours to make dipped candles. The process is fun, creative, and productive. Give your handmade candles to friends or family or burn them at home to create atmosphere and save on your electricity bill.

Rolled Beeswax Candle

Beeswax candles are cheap, eco-friendly, non-allergenic, dripless, non-toxic, and they burn cleanly and beautifully.

TIP

Rather than pouring leftover wax down the drain (which will clog your drain and is bad for the environment), dump it into a jar and set it aside. You can melt it again later for another project.

TIP

When making candles, keep a box of baking soda nearby. If wax lights on fire, it reacts similarly to a grease fire, which is aggravated by water. Douse a wax fire with baking soda and it will extinguish quickly.

They're also very simple and quick to make—perfect for a short afternoon project.

Materials

- Sheets of beeswax (you can find these at your local arts and crafts store or from a local beekeeper)
- Wick (you can purchase candle wicks at your local arts and crafts store)

Supplies

- Scissors
- Hair dryer (optional)

Directions

1. Fold one sheet of beeswax in half. Cut along the crease to make two separate pieces.
2. Cut your wick to about 2 inches longer than the length of the beeswax sheet.
3. Lay the wick on the edge of the beeswax sheet, closest to you. Make sure the wick hangs off of each end of the sheet.

TIP

If you are having trouble using the beeswax and want to facilitate the adhering process, you can use a hair dryer to soften the wax and to help you roll it. Start at the end with the wick and, moving the hair dryer over the wax, heat it up. Keep rolling until you reach a section that is not as warm, heat that up, and continue all the way to the end.

TIP

Old crayons can be melted and used instead of paraffin for candle-making.

4. Start rolling the beeswax over the wick. Apply slight pressure as you roll to keep the wax tightly bound. The tighter you roll the beeswax, the sturdier your candle will be and the better it will burn.
5. When you reach the end, seal off the candle by gently pressing the edge of the sheet into the rolled candle, letting your body heat melt the wax.
6. Trim the wick on the bottom (you may also want to slice off the bottom slightly to make it even so it will stand up straight) and then cut the wick to about ½ inch at the top.

Taper Candles

Taper candles are perfect for candlesticks, and they can be made in a variety of sizes and colors.

Materials

- Wick (be sure to find a spool of wick that is made specifically for taper candles)
- Wax (paraffin is best)
- Candle fragrances and dyes (optional)

Supplies

- Pencil or chopstick (to wind the wick around to facilitate dipping and drying)
- Weight (such as a fishing lure, bolt, or washer)
- Dipping container (this should be tall and skinny. You can find these containers at your local arts and craft store, or you can substitute a spaghetti pot)
- Stove
- Large pot for boiling water
- Small trivet or rack
- Glass or candle thermometer

LAYERED TAPER CANDLES

For a more ornate candle, add different shades of food coloring to three or four separate pots of melted wax (or melt down old crayons). Alternate between the different colors of wax as you dip the wick, creating different layers of color. Once the candle is the desired thickness and is mostly cooled, use a paring knife to carefully peel away strips of the wax around the outside of the candle. Allow the wax strips to curl downward as you peel, revealing a rainbow of colors.

- Newspaper
- Drying rack

Directions

1. Cut the wick to the desired length of your candle, leaving about 5 additional inches that will be tied onto the pencil or chopstick for dipping and drying purposes. Attach a weight (a fishing lure, bolt, or heavy metal washer) to the dipping end of the wick to help with the first few dips into the wax.
2. Ready your dipping container. Put the wax (preferably in smaller chunks—this will speed up the melting process) into the container and set aside.
3. In a large pot, start to boil water. Before putting the dipping container full of wax into the larger pot, place a small trivet, rack, or other elevating device into the bottom of the larger pot. This will keep the dipping container from touching the bottom of the larger pot and will prevent the wax from burning and possibly combusting.
4. Put the dipping container into the pot and start to melt the wax, keeping a thermometer in the wax at all times. The wax should be heated and melted between 150 and 165°F. Stir frequently in order to keep the chunks of paraffin from burning and to make sure all the wax is thoroughly melted. (If you want to add fragrance or dye, do so when the wax is completely melted and stir until the additives are dissolved.)
5. Once your wax is completely melted, it's time to start the dipping process. Removing the container from the stove, take your wick that's tied onto a stick and dip it into the wax, leaving it there for a few minutes. Continue to lower the wick in and out of the dipping container, and by the eighth or ninth dip, cut off the weight from the bottom of the wick—the candle should be heavy enough now to dip well on its own.

6. To speed up the cooling process—and to help the wax to continue to adhere and build up on the wick—blow on the hot wax each time you lift the candle out of the dipping pot.
7. When the candle is at the desired length and thickness, you may want to lay it down on a very smooth surface (such as a countertop) and gently roll it into shape.
8. On a drying rack (which can be made from a box long enough so the candles do not touch the bottom or from another device), carefully hang your taper candle to dry for a good twenty-four hours.
9. Once the candle is completely hardened, trim the wick to just above the wax.

TIP

To make floating candles, pour hot wax into a muffin tin until each muffin cup is about one-third full. Allow the wax to cool until a film forms over the tops of the candles. Insert a piece of wick into the center of each candle (use a toothpick to help poke the hole if necessary). Allow candles to finish hardening and then pop them out of the tin. Trim wicks to about ¼ inch.

GOURD VOTIVES

Small gourds make perfect votive candleholders. Carve a circle out of the top of the gourd, making it the same size as the circumference of the candle you intend to place in it. Gently pry off the top and set the candle in the indentation. If necessary, cut the hole slightly larger, but keep it small enough that the candle fits snugly.

Jarred Soy Candles

Soy candles are environmentally friendly and easy to make. You can find most of the ingredients and materials needed to make soy candles at your local arts and crafts store—or even in your own kitchen!

Materials

- 1 lb soy wax (either in bars or flakes)
- 1 ounce essential oil (for fragrance)
- Natural dye (try using dried and powdered beets for red, turmeric for yellow, or blueberries for blue)

Supplies

- Stove
- Pan to heat wax (a double boiler is best)
- Spoon
- Glass thermometer
- Candle wick (you can find this at your local arts and crafts store)
- Metal washers
- Pencils or chopsticks
- Heatproof cup to pour your melted wax into the jar(s)
- Jar to hold the candle (jelly jars or other glass jars work well)

TIP

Add citronella essential oil and a few drops of any of the following other essential oils to make your candle a mosquito repellant:

- Catnip
- Cloves
- Cedarwood
- Eucalyptus
- Lavender
- Lemongrass
- Peppermint
- Rose geranium
- Rosemary
- Thyme

Directions

1. Put the wax in a pan or a double boiler and heat it slowly over medium heat. Heat the wax to 130 to 140°F or until it's completely melted.
2. Remove the wax from the heat. Add the essential oil and dye (optional) and stir into the melted wax until completely dissolved.
3. Allow the wax to cool slightly, until it becomes cloudy.
4. While the wax is cooling, prepare your wick in the glass container. It is best to have a wick with a metal disk on the end—this will help stabilize it while the candle is hardening. If your wick does not already have a metal disk at the end, you can easily attach a thin metal washer to the end of the wick. Position the wick in the glass container and wrap the excess wick around the middle of a pen or chopstick. Lay the pencil or chopstick on the rim of the container and position the wick so it falls in the center.
5. Using a heat proof cup or the container from the double boiler, carefully pour the wax into the glass container, being careful not to disturb the wick from the center.
6. Allow the candle to dry for at least twenty-four hours before cutting off the excess wick and using.

Soap Making

When you make your own soap, you get to choose how you want it to look, feel, and smell. Adding dyes, essential oils, texture (with oatmeal, seeds, etc.), or pouring it into molds will make your soap unique. Making soap requires time, patience, and caution, as you'll be using some caustic and potentially dangerous ingredients—especially lye (sodium hydroxide). Avoid coming into direct contact with the lye; wear goggles, rubber gloves, and long sleeves, and work in a well-ventilated area. Be careful not to breathe in the fumes produced by the lye and water mixture.

Soap is made up of three main ingredients: water, lye, and fats or oils. While lard and tallow were once used exclusively for making soaps, it is perfectly acceptable to use a combination of pure oils for the "fat" needed to make soap. Saponification is the process in which the mixture becomes completely blended and the chemical reactions between the lye and the oils, over time, turn the mixture into a hardened bar of usable soap.

Cold-pressed Soap

Ingredients

- 6.9 ounces lye (sodium hydroxide)
- 2 cups distilled water, cold (from the refrigerator is the best)
- 2 cups canola oil
- 2 cups coconut oil
- 2 cups palm oil

Supplies

- Goggles, gloves, and mask (optional) to wear while making the soap
- Mold for the soap (a cake or bread loaf pan will work just fine; you can also find flexible plastic molds at your local arts and crafts store)
- Plastic wrap or wax paper to line the molds
- Glass bowl to mix the lye and water

- Wooden spoon for mixing
- 2 thermometers (one for the lye and water mixture and one for the oil mixture)
- Stainless steel or cast iron pot for heating oils and mixing in lye mixture
- Handheld stick blender (optional)

Directions

1. Put on the goggles and gloves and make sure you are working in a well-ventilated room.
2. Ready your mold(s) by lining with plastic wrap or wax paper. Set them aside.
3. Slowly add the lye to the cold, distilled water in a glass bowl (*never* add the water to the lye) and stir continually for at least a minute, or until the lye is completely dissolved. Place one thermometer into the glass bowl and allow the mixture to cool to around 110°F (the chemical reaction of the lye mixing with the water will cause it to heat up quickly at first).
4. While the lye is cooling, combine the oils in a pot on medium heat and stir well until they are melted together. Place a thermometer into the pot and allow the mixture to cool to 110°F.
5. Carefully pour the lye mixture into the oil mixture in a small, consistent stream, stirring continuously to make sure the lye and oils mix properly. Continue stirring, either by hand (which can take a very long time) or with a handheld stick blender, until the mixture traces (has the consistency of thin pudding). This may take anywhere from thirty to sixty minutes or more, so be patient. It is well worth the time invested to make sure your mixture traces. If it doesn't trace all the way, it will not saponify correctly and your soap will be ruined.
6. Once your mixture has traced, pour carefully into the mold(s) and let sit for a few hours. Then, when the mixture is still soft but congealed enough not to melt back into itself, cut the soap with a table knife into bars. Let sit for a few days, then take the bars out of the mold(s) and place on brown paper (grocery bags are perfect) in a dark area. Allow the bars to cure for another four weeks or so before using.

If you want your soap to be colored, add special soap-coloring dyes (you can find these at the local arts and crafts store) after the mixture has traced, stirring them in. Or try making your own dyes using herbs, flowers, or spices.

To make a yummy-smelling bar of soap, add a few drops of your favorite essential oils (such as lavender, lemon, or rose) after the tracing of the mixture and stir in. You can also add aloe and vitamin E at this point to make your soap softer and more moisturizing.

To add texture and exfoliating properties to your soap, you can stir some oats into the traced mixture, along with some almond essential oil or a dab of honey. This will not only give your soap a nice, pumice-like quality but it will also smell wonderful. Try adding bits of lavender, rose petals, or citrus peel to your soap for variety.

To make soap in different shapes, pour your mixture into molds instead of making them into bars. If you are looking to have round soaps, you can take a few bars of soap you've just made, place them into a resealable plastic bag, and warm them by putting the bag into hot water (120°F) for thirty minutes. Then, cut the bars up and roll them into balls. These soaps should set in about one hour or so.

SOAP OILS

Oil	Qualities
Almond Butter	Conditioning. Creamy Lather. Moderate Iodine.
Almond Oil, sweet	Conditioning. Fragrant. High Iiodine.
Apricot Kernel Oil	Conditioning. Fragrant. High Iodine.
Avocado Oil	Conditioning. Creamy Lather. High Iodine.
Babassu Oil	Cleansing. Bubbly. Very Low Iodine.
Canola Oil	Conditioning. Inexpensive. High Iodine.
Cocoa Butter	Creamy Lather. Low Iodine.
Coconut Oil	Bubbly Lather. Cleansing. Low Iodine.
Emu Oil	Conditioning. Creamy Lather. Moderate Iodine.
Evening Primrose Oil	Conditioning. Very High Iodine.
Flax Oil, Linseed	Conditioning. Very High Iodine.
Ghee	Cleansing. Bubbly Lather. Very Low Iodine.
Grapeseed Oil	Conditioning. Very High Iodine.
Hemp Oil	Conditioning. Very High Iodine.
Lanolin Liquid Wax	Low Iodine.
Neem Tree Oil	Conditioning. Creamy Lather. High Iodine.
Olive Oil	Conditioning. Creamy Lather. High Iodine.
Palm Oil	Conditioning. Creamy Lather. Moderate Iodine.
Rapeseed Oil	High Iodine.
Safflower Oil	Conditioning. Very High Iodine.
Sesame Oil	Conditioning. High Iodine.
Shea Butter	Conditioning. Creamy Lather. Moderate Iodine.
Ucuuba Butter	Conditioning. Creamy Lather. Low Iodine.

THE JUNIOR HOMESTEADER

How Soap Works

Teach kids how soap cleans with this simple experiment.

1. Half fill two Mason jars with water and add a few drops of food coloring. Pour several tablespoons of oil into each jar (corn oil, olive oil, or whatever you have on hand will be fine). You will see that the oil and water form separate layers. This is because the molecules in oil are hydrophobic, meaning that they repel water.
2. Add a few drops of liquid soap to one of the jars. Close both jars securely and shake for about thirty seconds. The oil and water should be thoroughly mixed.
3. Let both jars rest undisturbed. The jar with the soap in it will stay mixed, whereas the jar without the soap will separate back into two distinct layers. Why? Soap is made up of long molecules, each with a hydrophobic end and a hydrophilic (water-loving) end. The water bonds with the hydrophilic end and the oil bonds with the hydrophobic end. The soap serves as a glue that sticks the oil and water together. When you rinse off the soap, it sticks to the water, and the oil sticks to the soap, pulling all the oil down the drain.

NATURAL BATH SALTS AND SCRUBS

Follow these recipes to make your own luxurious bath products.

Lavender Bath Salt

Pour several tablespoons of this into your bath as it fills for an extra-soothing, relaxing, and cleansing experience. You can also add powdered milk or finely ground, old-fashioned oatmeal to make your skin especially soft. Toss in a few lavender buds if you have them.

Ingredients

- 2 cups coarse sea salt
- ½ cup Epsom salts
- ½ cup baking soda
- 4 to 6 drops lavender essential oil
- Red and blue food coloring, if desired (use more red than blue to achieve a lavender color)

Mix all ingredients thoroughly and store in a glass jar or other airtight container.

Citrus Scrub

Use this invigorating scrub to wake up your senses in the morning. The vitamin C in oranges serves as an astringent, making it especially good for oily skin.

Ingredients

- ½ orange or grapefruit
- 3 tbsps cornmeal
- 2 tbsps Epsom salts or coarse sea salt

Squeeze citrus juice and pulp into a bowl and add cornmeal and salts to form a paste. Rub gently over entire body and then rinse.

Healing Bath Soak

This bath soak will relax tired muscles, help to calm nerves, and leave skin soft and fragrant. You may also wish to add blackberry, raspberry, or violet leaves. Dried or fresh herbs can be used.

- 2 tbsps comfrey leaves
- 1 tbsp lavender
- 1 tbsp evening primrose flowers
- 1 tsp orange peel, thinly sliced or grated
- 2 tbsps oatmeal

Combine herbs and tie up in a small muslin or cheesecloth sack. Leave under faucet as the tub fills with hot water. If desired, empty herbs into the bath water once the tub is full.

Rosemary Peppermint Foot Scrub

Use this foot rub to remove calluses, soften skin, and leave your feet feeling and smelling wonderful.

Ingredients

- 1 cup coarse sea salt
- ¼ cup sweet almond or olive oil
- 2 to 3 drops peppermint essential oil
- 1 to 2 drops rosemary essential oil
- 2 sprigs fresh rosemary, crushed, or ½ tsp dried rosemary

Combine all ingredients and massage into feet and ankles. Rinse with warm water and follow with a moisturizer.

NATURAL DYES FOR SOAP OR CANDLES	
Light/Dark Brown	Cinnamon, ground cloves, allspice, nutmeg, coffee
Yellow	Turmeric, saffron, calendula petals
Green	Liquid chlorophyll, alfalfa, cucumber, sage, nettles
Red	Annatto extract, beets, grapeskin extract
Blue	Red cabbage
Purple	Alkanet root

Almost any oil can be used to make soap, but different oils have different qualities; some oils create a creamier lather, some create a bubbly lather. Oils that are high in iodine will produce a softer soap, so be sure to mix with oils that are lower in iodine. Online soap calculators are very helpful when creating your own recipes.

Cornhusk Dolls

This old-fashioned doll makes a wonderful gift for young children and also a unique, decorative, homemade item for your home or for sale at a craft fair. Cornhusk dolls are quite easy to make if you just follow these simple steps. You'll need:

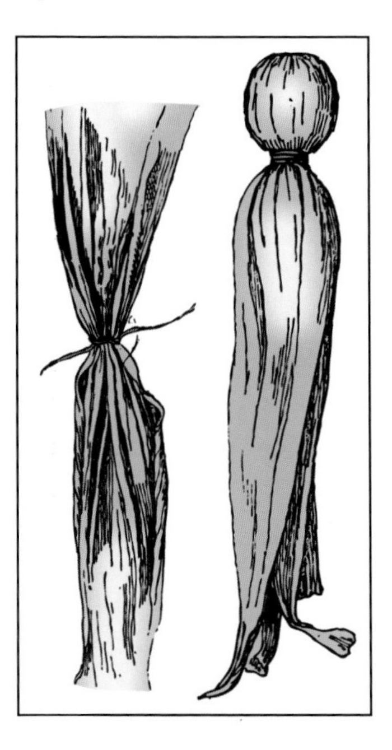

- Corn husks
- Thread or string
- Pen or marker
- Natural materials for decorations

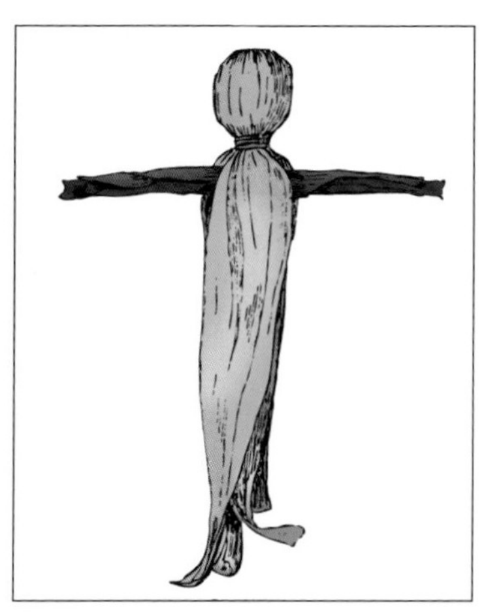

1. Gather husks from several large ears of corn (you may have these from your garden, if you grow corn, or you might find them at a garden center or a farmers' market in the fall). Select the soft, white husks that grow closest to the ear.

2. Place the stiff ends of two husks together, fold one of the long, soft husks in a strip lengthwise, and wrap it around the ends.
3. Choose the softest and widest husk you can find, fold it across the center, and place a piece of strong thread or string around it and tie it tightly in a knot.
4. Bring this down over the already-wound husks and tie it with a thread underneath. This will form the head and neck of the doll.
5. To make the arms, divide the husk below the neck into two equal parts. Fold two or three husks together and insert them in the space you've made by the division. Hold the arms in place with one hand and use your other hand to fold several layers of husk over each shoulder, allowing them to extend down the back of the figure.
6. When the figure seems substantial enough, use your best husks for the topmost layers and wrap the waist with strong thread, tying it tightly.
7. Divide the husks below the waist band and make the legs by neatly wrapping each portion with thread. Trim the husks off evenly for the feet.
8. Twist the arms once or twice, tie them, and trim them evenly for the hands.
9. You can draw a face on the doll with a pen or marker or you can glue tiny natural items on the head to make a face.
10. If you want your doll to have clothing or a specific "costume," you can make these from any kind of material and in any way you wish.

For a variation on a traditional cornhusk doll, try the following:

1. Gather a young ear of corn (whose silk has not yet browned), a crab apple for the head, and a leaf from the cob for a dress.

2. Cut off the bottom of the ear of corn where the husks are puckered and carefully take the silk from the other end, making sure not to disturb the closely wrapped husks remaining.
3. Roll part of the leaf (see picture above) for the arms and fasten the crab apple to the leaf arms with a small stick. Stick the other end of the twig into the small end of the corncob.
4. Now you can dress the doll. The hat for the doll can be made from a leaf (just where it joins the stalk). This can be fastened to the doll's head with a small twig or thorn. Make sure the silk is placed on the head to form hair before securing the hat.

BIRCH BARK BASKET

Birch bark baskets are a wonderful way to display dried wildflowers and make nice gifts. Making a basket in the shape of a canoe works very well with the bark. Gather bark strips (do not string directly from the tree; try to find these either on cut wood or from a craft or lumber store) that are 6½ inches long and 4 inches wide. Sew the ends of the bark together with a thick thread, leaving one side of each strip unstitched. Sew a ribbon on each end of the canoe—these will serve as handles. Now the basket may be filled with dried wildflowers or other things from nature (such as pine cones) and hung on the wall for display.

5. Make a scarf by folding a leaf around the shoulders and securing it with small pins or thorns (see picture on previous page).
6. Stick tiny thorns into the crab apple to make eyes and a nose.

Lampshades

Making your own lampshades will enable you to match or complement the décor of your room. Choose material that is fire-resistant; a 100-watt light bulb can heat up to more than 200 degrees. Keep in mind that lighter-colored, thin fabrics will let more light through than dark, heavy fabrics.

- Newspaper
- Lampshade frame
- Fabric
- Pins
- Scissors
- Masking tape
- Measuring tape or ruler
- Fabric glue

1. Measure the circumference of the top and bottom of your lampshade frame. To facilitate this, you can place a strip of masking tape around one circle and then peel it off and use a ruler or measuring tape to determine the exact length of the tape. Next, measure the distance from the bottom, wide circle of the frame to the top, narrow circle. Add 1 ½ inches to all measurements.
2. Use your measurements to draw a pattern on newspaper. The pattern will be in the shape of a trapezoid. The bottom of the trapezoid is the length of the bottom circle of the frame; the top of the trapezoid is the length of the top circle of the frame; and the distance between them is the height of your frame.
3. Cut out the pattern and lay it on the backside of your fabric. Pin the pattern to the fabric and carefully cut out the fabric. Remove the pattern.

4. Cover the backside of the fabric with fabric glue or spray adhesive. Place the frame onto the fabric and begin rolling it, pressing the fabric to the frame as you go. The fabric should hang over the top and bottom of the frame slightly.
5. Once the fabric is wrapped all the way around the lampshade, tuck in the raw edge of the fabric to create a seam and glue it down. Then glue and tuck the fabric over the top circle and under the bottom circle of the frame. Use clothespins to hold the fabric in place until it dries completely.
6. If desired, add ribbons or tassels to the top or bottom of the shade.

TIP

Fabric with a higher percentage of cotton will adhere to the frame better than synthetic fabrics. Also keep in mind that the darker the fabric, the less light will shine through.

Winter

Winter is perhaps the ideal season for crafting. The colder weather lends itself to long afternoons by the fire spent piecing together a quilt or organizing photos and keepsakes into a scrapbook. Or try your hand at making pottery or designing jewelry!

Quilting

Crazy quilts first became popular in the 1800s and were often hung as decorative pieces or displayed as keepsakes, but they can also be warm and practical. They can be made out of scraps of fabric that are too small for almost any other use, and there is endless room for variation in colors, patterns, and texture.

Quilts are generally made up of many small squares that are sewn together into a large rectangle and layered with batting (a thick layer of fabric to add warmth—usually wool or cotton) and backing (the material that will show on the underside of the quilt).

- Non-woven interfacing or lightweight muslin (prewashed)
- Fabric pieces in a variety of colors and patterns
- Scissors

- Ruler
- Pencil
- Needle or sewing machine
- Pins
- Thread
- Iron

1. Make the foundation squares. If your foundation fabric (interfacing or muslin) is wrinkled, iron it carefully until it is completely flat. Then use a ruler to measure and draw a 13 x 13-inch square in one corner (your final square will be 12 x 12 inches, but it's a good idea to leave yourself a little extra fabric to work with). Repeat until all of the foundation fabric is cut into squares.
2. Cut a small piece of patterned fabric into a shape with three or five straight edges. Pin it right side up on the center of one foundation square. Cut another small piece of fabric with straight edges and lay it right side down on top of the first piece. Sew a ¼-inch seam along the edge where the two fabrics overlap. If the second piece is longer than the first piece, don't sew beyond the edge of the first piece. Turn the second piece of fabric over so that it's facing up and iron it. Trim the second piece to align with the first, so you have one larger shape with straight edges.
3. Continue with a third piece of fabric, making sure it is large enough to extend the length of the first two patches combined. Sew the seam, flip the fabric upright, iron, trim, and proceed with a fourth piece. Work clockwise, each piece getting larger as you move toward the edge of the foundation square. Once the square is filled, trim off any overhanging fabric so that you have one neat square. Repeat steps 2 and 3 until all foundation pieces are filled.
4. Sew all foundation squares together with a ¼- to ½-inch seam.
5. Sandwich the quilt by placing the backing face down, the batting on top of it, and then the foundation on top, with the patterned squares facing up. Baste around the quilt to hold the three layers together, using long stitches and staying about ¼-inch from the outside edge of the patterned fabric.
6. Binding your quilt covers the rough edges and creates an attractive border around the edges of the quilt. To make the binding, cut strips of fabric 2 ½ inches wide and as long as one side of your quilt plus 2 inches. Fold the fabric in half lengthwise and press.
7. Lay the strip along one edge of the quilt. The raw edges of the quilt and the binding should be stacked together. Leave a ½ inch extra hanging off the first corner. Sew along the length of the quilt, about ¼ inch from the raw edge. Trim the binding, leaving a ½ inch extra. Fold the fabric over the rough edges to the back of the quilt and slip stitch the binding to the backside. Fold the loose ends of the binding over the edge of the quilt and stitch to the backside. Repeat with all sides of the quilt.
8. Finish your crazy quilt by adding decorative stitching between small pieces of fabric, sewing on buttons, tassels, or ribbons, or using stitching or fabric markers to record important names or dates.

These instructions are for making pottery for decorative use. People wanting to make pottery for functional use should seek the help of a studio with a kiln and food safe glazes.

Pottery Basics

Clay is the basic ingredient for making pottery. Clay is decomposed rock containing water (both in liquid and chemical forms). Water in its liquid form can be separated from the clay by heating the mass to a boiling point—a process that restores the clay to its original condition once dried. The water in the clay that is found in chemical forms can also be removed by ignition—a process commonly referred to as "firing." After being fired, clay cannot be restored to any state of plasticity—this is called "pottery." Some clay requires greater heat in order to be fired, and these are known as "hard clays." These types of clay must be subjected to a "hard-firing" process. However, in the making of simple pottery, soft clay is generally used and is fired in an overglaze (soft glaze) kiln.

Pottery clays can either be made by hand (by finding clay in certain soils) or bought from craft stores. If you have clay soil available on your property, the process of separating the clay from the other soil materials is simple. Put the earthen clay into a large bucket of water to wash the soil away. Any rocks or other heavy matter will sink to the bottom of the bucket. The milky fluid that remains—which is essentially water mixed with clay—may then be drawn off and allowed to settle in a separate container, the clear water eventually collecting on the top. Remove the excess water by using a siphon. A repetition of this process will refine the clay and make it ready for use.

You can also purchase clay at your local craft store. Usually, clay sold in these stores will be in a dry form (a grayish or yellowish powder), so you will need to prepare it in order to use it in your pottery. To prepare it for use, you must mix the powder with water. If there are directions on your clay packet, then follow those closely to make your clay. In general, though, you can make your clay by mixing equal parts of clay powder and water in a bowl and allowing the mixture to soak for ten to twelve hours. After it has soaked, you must knead the mixture thoroughly to disperse the water evenly throughout the clay and pop any air bubbles. Air bubbles, if left in the clay, could be detrimental to your pottery once kilned, as the bubbles would generate steam and possibly crack your creation. However, be careful not to knead your clay mixture too much, or you may increase the chance of air bubbles becoming trapped in the mixture.

If, after kneading, you find that the clay is too wet to work with (test the wetness of the clay on your hands and if it tends to slip around your palm very easily, it is probably too wet), the excess water can be removed by squeezing or blotting out with a dry towel or dry-board.

The main tools needed for making pottery are simply your fingers. There are wooden tools that can be used for adding finer detail or decoration, but typically, all you really need are your own two hands. A loop tool (a piece of fine, curved wire) may also be used for scraping off excess clay where it is too thick. Another tool has ragged edges and this can be used to help regulate the contour of the pottery. Remember that homemade pottery will not always be symmetrical, and that is what makes it so special.

Basic Vase or Urn

Try making this simple vase or urn to get used to working with clay.

1. Take a lump of clay. The clay should be about the size of a small orange and should be rather elastic feeling. Then, begin to mold the base of your object—let's say it is either a bowl or a vase.
2. Continue molding your base. By now, you'll have a rather heavy and thick model, hollowed to look a little like a bird's nest. Now, using this base as support, start adding pieces of clay in a spiral shape. Press the clay together firmly with your fingers. Make sure that your model has a uniform thickness all around.
3. Continue molding your clay and making it grow. As you work with the clay, your hands will become more accustomed to its texture and the way it molds, and you will have less difficulty making it do what you want. As you start to elongate and shape the model, remember to keep the walls of the piece substantial and not too thin—it is easier to remove extra thickness than it is to add it.
4. Don't become frustrated if your first model fails. Even if you are being extra careful to make your bowl or vase sturdy, there is always the instance when a nearly complete vase will fall over. This usually happens when one side of the structure becomes too thin or the clay is too wet. To keep this from happening, it is sometimes helpful to keep one hand inside the structure and the other outside. If you are building a vase, you can extract one finger at a time as you reach closer and closer to the top of the model.
5. Make sure the clay is moist throughout the entire molding process. If you need to stop molding for an extended period of time, cover the item with a moist cloth to keep it from drying out.
6. When your model has reached the size you want, you may turn it upside down and smooth and refine the contours of the object. You can also make the base much more detailed and shaped to a more pleasing design.
7. Allow your model to air dry.

Embellishing Your Clay Models

You may eventually want to make something that requires a handle or a spout, such as a cup or teapot. Adding handles and spouts can be tricky, but only if you don't remember some simple rules. Spouts can be modeled around a straw or any other material that is stiff enough to support the clay and light enough to burn out in the firing. In the designing of spouts and handles, it is still important to keep them solid and thick. Also, keeping them closer to the body of your model is more practical, as handles and spouts that are elongated are harder to keep firm and can also break off easily. Although more time consuming and difficult to manage, handles and spouts can add a nice aesthetic touch to your finished pottery.

The simplest way to decorate your pottery is by making line incisions. Line incision designs are best made with wooden, finger-shaped tools. It is completely up to you as to how deep the lines are and into what pattern they are made.

Wheel-working and Firing Pottery

If you want to take your pottery making one step further, you can experiment with using a potters' wheel and also glazing and firing your model to create beautiful pottery. Look online or at your local craft store for potters' wheels. Firing can leave your pottery looking two different ways, depending on whether you decide to leave the clay natural (so it maintains a dull and porous look) or to give it a color glaze.

Colored glazes come in the form of powder and are generally metallic oxides, such as iron oxides, cobalt oxide, chromium oxide, copper oxide, and copper carbonate. The colors these compounds become will vary depending on the atmosphere and temperature of the kiln. Glazes often come in the form of powder and need to be combined with water in order to be applied to the clay. Only apply glaze to dried pottery, as it won't adhere well to wet clay. Use a brush, sponge, or putty knife to apply the glaze. Your pottery is then ready to be fired.

There are various different kinds of kilns in which to fire your pottery. An overglaze kiln is sufficient for all processes discussed here, and you can probably find a kiln in your surrounding area (check online and in your telephone book for places that have kilns open to the public). Schools that have pottery classes may have overglaze kilns. It is important, whenever you are using a kiln, that you are with a skilled pottery maker who knows how to properly operate the kiln.

After the pottery has been colored and fired, a simple design may be made on the pottery by scraping off the surface color so as to expose the original or creamy-white tint of the clay.

Unglazed pottery may be worked with after firing by rubbing floor wax on the outer surface. This fills up the pores and gives a more uniform quality to the whole piece.

Pottery offers so many opportunities for personal experimentation and enjoyment; there are no set rules as to how to make a piece of pottery. Keep a journal about the different things you try while making pottery, so you can remember what works best and what should be avoided in the future. Note the kind of clay you used and its consistency, the types of colors that have worked well, and the temperature and positioning within the kiln, if you use firing. Above all, enjoy making unique pieces of pottery!

Making Jars, Candlesticks, Bowls, and Vases

Making pottery at home is simple and easy, and is a great way for you to make personalized, unique gifts for family and friends. Clay can be purchased at local arts and crafts stores. Clay must always be kneaded before you model with it, because it contains air that, if left in the clay, would form air bubbles in your pottery and spoil it. Work out this air by kneading it the same way that you knead bread. Also guard against making the clay too moist, because that causes the pottery to sag, which spoils the shape.

In order to make your own pottery, you need modeling clay, a board on which you can work, a pie tin on which to build, a knife, a short stick (one side should be pointed), and a ruler.

Jars

To start a jar, put a handful of clay on the board, pat it out with your hand until it is an inch thick, and smooth off the surface. Then, take a coffee cup, invert it upon the base, and, with your stick, trim the clay outside the rim.

To build up the walls, put a handful of clay on the board and use a knife to smooth it out into a long piece, ¼ inch thick. With the knife and a ruler, trim off one edge of the piece and cut a number of strips ¾ inch wide. Take one strip, stand it on top of the base, and rub its edge into the base on both sides of the strip; then, take another strip and add it to the top of the first one, and continue building in this way, placing one strip on another, joining each to the one beneath it, and smoothing over the joints as you build. Keep doing this until the walls are as high as you want them to be. Remember to keep one hand inside the jar while you build, for extra support. Fill uneven places with bits of clay and smooth out rough spots with your fingers, having moistened your fingers with water first. When you are finished, you may also add decoration, or ornament, to your jar.

Candlesticks

Making a pottery candlestick requires a round base ½ inch thick and 4 inches in diameter. After preparing the base, put a lump of clay in the center, work it into the base, place another lump on top, work it into the piece, and continue in this way until the candlestick has been built as high as you want it. Then, force a candle into the moist clay, twisting it around until it has made a socket deep enough to place a candle into.

A cardboard "templet," with one edge trimmed to the proper shape, will help make it easy to make the walls of the candlestick symmetrical and the projecting cap on the top equal on all sides. Run the edge of the

templet around the walls as you work, and it will show you exactly where and how much to fill out, trim, and straighten the clay.

If you want to make a candlestick with a handle, make a base just as stated above. Then cut strips of clay and build up the wall as if building a jar, leaving a center hole just large enough to hold a candle. When the desired height for the wall has been reached, cut a strip of clay ½ inch wide and ½ inch thick, and lay it around the top of the wall with a projection of ¼ inch over the wall. Smooth this piece on top, inside, and outside with your modeling stick and fingers. For the handle, prepare a strip 1 inch wide and ½ inch thick, and join one end to the top band and the other end to the base. Use a small lump of clay for filling around where you join the piece, and smooth off the piece on all sides.

When the candlestick is finished, run a round stick the same size as the candle down into the hole, and let it stay put until the clay is dry, to keep the candlestick straight.

Bowls

Bowls are simple to make. Starting with a base, lay strips of clay around the base, building upon each strip as you did when making a jar. Once the bowl is at the desired height and width, allow it to dry.

Vases

Vases can be made as you'd make a jar, only bringing the walls up higher, like a candlestick. Experiment with different shapes and sizes and always remember to keep one hand inside the vase to keep everything even and to prevent your vase from caving in on itself.

Decorating Your Pottery

Pottery may be ornamented by scratching a design upon it with the end of your modeling stick. You can do a simple, straight-line design by using a ruler to guide the stick in drawing the lines. Ornamentation on vases and candlesticks can be done by hand-modeling details and applying them to your item.

Glazing and Firing

Pottery that you buy is generally glazed and then fired in a pottery kiln, but firing is not necessary to make beautiful, sturdy pottery. The clay will dry hard enough, naturally, to keep its shape, and the only thing you must provide for is waterproofing (if the pottery will be holding liquids). To do this, you can take bathtub enamel and apply it to the inside (and outside, if desired) of the pottery to seal off any cracks and to keep the item from leaking.

If you do want to try glazing and firing your own pottery, you will need a kiln. Below are instructions for making your own.

Sawdust Kiln

This small, homemade kiln can be used to bake and fire most small pottery projects. It will only get up to about 1,200°F, which is not hot enough to fire porcelain or stoneware. However, it will suffice for clay pinch pots and other decorative pieces.

You will need:

- 20 to 30 red or orange bricks
- Chicken wire
- Sawdust
- Newspaper and kindling
- Sheet metal

1. Choose a spot outdoors that is protected from strong winds. Clear away any dried

branches or other flammables from the immediate area. A concrete patio or paved area makes an ideal base, but you can also place bricks or stones on the ground.

2. Stack bricks in a square shape, building each wall up at least four bricks high. Fill the kiln with sawdust.
3. Place the chicken wire on top of the bricks and add another layer or two of bricks. Carefully place your pottery in the center of the mesh, spacing the pieces at least ½ inch apart. Cover the pottery with sawdust.
4. Add another piece of chicken wire, add bricks and pottery, and cover with sawdust. Repeat until your kiln is the desired height.
5. Light the top layer of sawdust on fire, using kindling and newspaper if needed. Cover with the sheet metal, using another layer of bricks to hold it in place.
6. Once the kiln stops smoking, leave it alone until it completely cools down. Then carefully remove the sheet metal lid.

Handcrafted Paper

Instead of throwing away your old newspapers, office paper, or wrapping paper, use it to make your own unique paper! The paper will be much thicker and rougher than regular paper, but it makes great stationery, gift cards, and gift wrap.

Materials

- Newspaper (without any color pictures or ads if at all possible), scrap paper, or wrapping paper (non-shiny paper is preferable)
- 2 cups hot water for every ½ cup shredded paper
- 2 tsps instant starch (optional)

Supplies

- Blender or egg beater
- Mixing bowl
- Flat dish or pan (a 9 x 13-inch or larger pan will do nicely)
- Rolling pin
- 8 x 12-inch piece of non-rust screen
- 4 pieces of cloth or felt to use as blotting paper, or at least 1 sheet of formica
- 10 pieces of newspaper for blotting

Directions

1. Tear the newspaper, scrap paper, or wrapping paper into small scraps. Add hot water to the scraps in a blender or large mixing bowl.
2. Beat the paper and water in a blender or with an egg beater in a large bowl. If you want, mix in the instant starch (this will make the paper ready for ink). The paper pulp should be the consistency of a creamy soup when it is complete.
3. Pour the pulp into the flat pan or dish. Slide the screen into the bottom of the pan. Move the screen around in the pulp until it is evenly covered.

4. Carefully lift the screen out of the pan. Hold it level and let the excess water drip out of the pulp for a minute or two.
5. With the pulp side up, put the screen on a blotter (felt) that is situated on top of some newspaper. Put another blotter on the top of the pulp and put more newspaper on top of that.
6. Using the rolling pin, gently roll the pin over the blotters to squeeze out the excess water. If you find that the newspaper on the top and bottom is becoming completely saturated, add more (carefully) and keep rolling.
7. Remove the top level of newspaper. Gently flip the blotter and the screen over. Very carefully, pull the screen off of the paper. Leave the paper to dry on the blotter for at least twelve to twenty-four hours. Once dry, peel the paper off the blotter.

To add variety to your homemade paper:

- To make colored paper, add a little bit of food coloring or natural dye to the pulp while you are mixing in the blender or with the egg beater.
- You can also try adding dried flowers (the smoother and flatter, the better) and leaves or glitter to the pulp.
- To make unique bookmarks, add some small seeds to your pulp (hardy plant seeds are ideal), make the paper as in the directions, and then dry your paper quickly using a hairdryer. When the paper is completely dry, cut out bookmark shapes and give to your friends and family. After they are finished using the bookmarks, they can plant them and watch the seeds sprout.

Scrapbooking

Creating a scrapbook is a great way to process and preserve memories. Keep travel brochures from special trips, ticket stubs, event programs, photographs, and so on in a box or file throughout the year. Then, when cold or rainy weather sets in, you can pull out

your book and begin arranging your keepsakes in a meaningful way.

Depending on the nature of your keepsakes and your own preferences, you can choose to base your scrapbook on a theme (musical or sporting events, family vacations, your wedding, home renovations, etc.) or to organize it chronologically. Of course, really you can do anything you want, including creating a collage-like book of scattered memories, keepsakes, magazine clippings, or even journal entries. Scrapbooking is a personal project and can be done with as much thoughtful organization or creative chaos as you wish.

There are many ways to create a unified feel to your scrapbook, whether or not the subject matter itself maintains any continuity. Here are a few suggestions:

- Layout. Design each page in a similar way, so that (for example) there is always a favorite quote on the top of the page, a photo on the side, text below, and an embellishment across from it. Or create a pattern of layouts, so that every third or fourth page has a similar design.
- Colors. Choose two or three colors or a single color palette to use throughout the book. Incorporate these colors in borders, frames for photos, backgrounds, with colored inks, and so on.
- Fonts or Script. Use the same style of writing for photo captions or other text throughout the book.
- Backgrounds. You may want a clean white background if you want to showcase high-quality photographs without any distractions to the eye. Or you can use scrap pieces of wallpaper all in the same style or color, colored card stock, or even thin fabric as the backdrop for your keepsakes.

Once you have a general plan for your scrapbook, you're ready to begin:

1. Create a background. Unless you want a plain white background, select the cardstock or other colored papers or fabrics, cut them to size, and use rubber cement to glue them to the page. You may choose to use several coordinated pieces of paper to create a patterned background.
2. Matt your photos. Cut squares of paper that are slightly larger than your photos and glue the photos onto them. This will create an attractive border around your photos. Later you can add frames or leave them as they are.
3. Add text or other embellishments. Use neat handwriting or printed text to caption your photos, add favorite quotes, etc. Pieces of ribbon, stickers, or magazine clippings can also be added as accents. Avoid any thick materials that will make your page bumpy and keep the album from closing all the way.

Rag Rugs

The first rag rugs were made by homesteaders over two centuries ago who couldn't afford to waste a scrap of fabric. Torn garments or scraps of leftover material could easily be turned into a sturdy rug to cover dirt floors or stave off the cold of a bare wooden floor in winter. Any material can be used for these rugs, but cotton or wool fabrics are traditional.

- Rags or strips of fabric
- Darning needle
- Heavy thread

TIP

Use thinner strips of fabric toward the end of your rug to make it easier to tack to the edge of the rug.

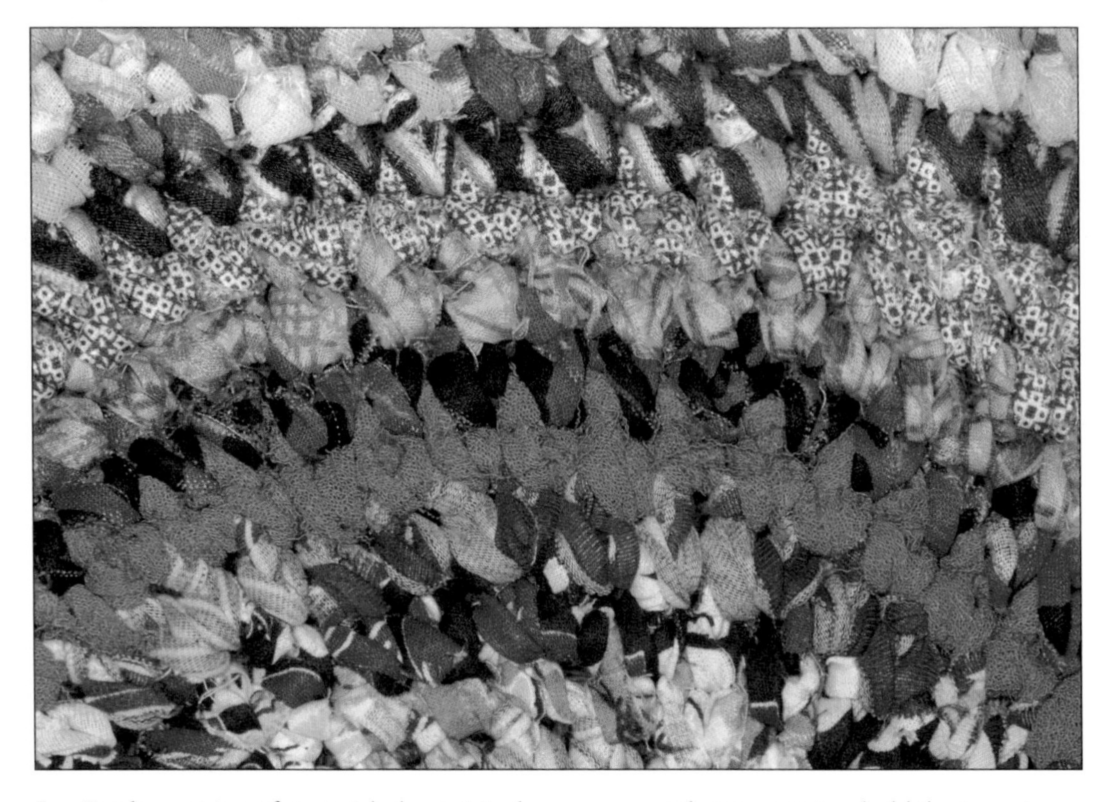

1. Cut long strips of material about 1 inch wide. Sew strips together end to end to make three very long strips (or you can start with shorter strips and sew on more pieces later). To make a clean seam between strips, hold the two pieces together at right angles to form a square corner. Sew diagonally across the square and trim off excess fabric.
2. Braid the three strips together tightly, just as you would braid hair.
3. Start with one end of the braid and begin coiling it around itself, sewing each coil to the one before it with circular stitches. Keep the coil flat on the floor or on a table to keep it from bunching up.
4. When the rug is as large as you want it to be, tack the end of the braid firmly to the edge of the rug.

≫ Cut strips along the bias to keep them from unraveling.

Part 4 The Barnyard

Chickens 208
Ducks 212
Turkeys 216
Rabbits 220
Beekeeping 223
Goats 228
Sheep 233
Llamas 237
Cows 241
Pigs 245
Butchering 249

Agriculture is our wisest pursuit, because it will in the end contribute most to real wealth, good morals, and happiness.

—Letter from Thomas Jefferson to George Washington (1787)

Chickens

Raising chickens in your yard will give you access to fresh eggs and meat, and since chickens are some of the easiest creatures to keep, even families in urban areas are able to raise a few in a small backyard. Four or five chickens will supply your whole family with eggs on a regular basis.

Housing Your Chickens

You will need to have a structure for your chickens to live in—to protect them from predators and inclement weather, and to allow the hens a safe place to lay their eggs.

Placing your henhouse close enough to your own home will remind you to visit it frequently to feed the chickens and to gather eggs. It is best to establish the house and yard in dry soil and to stay away from areas in your yard that are frequently damp or moist, as this is the perfect breeding ground for poultry diseases. The henhouse should be well-ventilated, warm, protected from the cold and rain, have a few windows that allow sunlight to shine in (especially if you live in a colder climate), and have a sound roof.

The perches in your henhouse should not be more than 2½ feet above the floor, and you should place a smooth platform under the perches to catch the droppings so they can easily be cleaned. Nesting boxes should be kept in a darker part of the house and should have ample space around them.

Chickens that are allowed to roam freely ("free-range" chickens) will be able to scavenge most of the food they need, as long as there is plenty of grass or other vegetation available.

The perches in your henhouse can be relatively narrow and shouldn't be more than a few feet from the floor.

A simple, movable chicken coop can be constructed out of two-by-fours and two wheels. The floor of the coop should have open slats so that the manure will fall onto the ground and fertilize the soil. An even simpler method is to construct a pen that sits directly on the ground, making sure that it has a roof to offer the chickens suitable shade. The pen can be moved once the area is well fertilized.

Selecting the Right Breed of Chicken

Take the time to select chickens that are well suited for your needs. If you want chickens solely for their eggs, look for chickens that are good egg-layers. Mediterranean poultry are good for first-time chicken owners as they are easy to care for and only need the proper food in order to lay many eggs. If you are looking to slaughter and eat your chickens, you will want to have heavy-bodied fowl (Asiatic poultry) in order to get the most meat from them. If you are looking to have chickens that lay a good amount of eggs and that can also be used for meat, invest in the Wyandottes or Plymouth Rock breeds. These chickens are not incredibly bulky but they are good sources of both eggs and meat.

Wyandottes have seven distinct breeds: Silver, White, Buff, Golden, and Black are the most common. These breeds are hardy and they are very popular in the United States. They are compactly built and lay excellent dark brown eggs. They are good sitters and their meat is perfect for broiling or roasting.

Plymouth Rock chickens have three distinct breeds: Barred, White, and Buff. They are the most popular breeds in the United States and are hardy birds that grow to a medium size. These chickens are good for laying eggs, roost well, and also provide good meat.

Building a chicken coop close to your house will make it easier to tend the chickens and gather eggs in inclement weather.

Feeding Your Chickens

Chickens, like most creatures, need a balanced diet of protein, carbohydrates, vitamins, fats, minerals, and water. Chickens with plenty of access to grassy areas will find most of what they need on their own. However, if you don't have the space to allow your chickens to roam free, commercial chicken feed is readily available in the form of mash, crumbles, pellets, or scratch. Or you can make your own feed out of a combination of grains, seeds, meat scraps or protein-rich legumes, and a gritty substance such as bone meal, limestone, oyster shell, or granite (to aid digestion, especially in winter). The correct ratio of food for a warm, secure chicken should be 1 part protein to 4 parts carbohydrates. Do not rely too heavily on corn as it can be too fattening for hens; combine corn with wheat or oats for the carbohydrate portion of the feed. Clover and other green foods are also beneficial to feed your chickens.

How much food your chickens need will depend on breed, age, the season, and how much room they have to exercise. Often it's easiest and best for the chickens to leave feed available at all times in several locations within the chickens' range. This will ensure that even the lowest chickens in the pecking order get the feed they need.

Hatching Chicks

If you are looking to increase the number of chickens you have, or if you plan to sell some chickens at the market, you may want some of your hens to lay eggs and hatch chicks. In order to hatch a chick, an egg must be incubated for a sufficient amount of time with the proper heat, moisture, and position. The period for incubation varies based on the species of chicken. The average incubation period is around twenty-one days for most common breeds.

If you are only housing a few chickens in your backyard, natural incubation is the easiest method with which to hatch chicks. Natural incubation is dependent upon the instinct of the mother hen and the breed of hen. Plymouth Rocks and Wyandottes are good hens to raise chicks. It is important to separate the setting hen from the other chickens while she is nesting and to also keep the hen clean and free from lice. The nest should also be kept clean and the hens should be fed grain food, grit, and clean, fresh water.

It is important, when you are considering hatching chicks, to make sure your hens are healthy, have plenty of exercise, and are fed a balanced diet. They need materials on which to scratch and should not be infested with lice and other parasites. Free range chickens, which eat primarily natural foods and get lots of exercise, lay more fertile eggs than do tightly confined hens. The eggs selected for hatching should not be more than twelve days old and should be clean.

You'll need to construct a nesting box for the roosting hen and the incubated eggs. The box should be roomy and deep enough to retain the nesting material. Treat the box with a disinfectant before use to keep out lice, mice, and other creatures that could infect

Chicken Feed

- 4 parts corn (or more in cold months)
- 3 parts oat groats
- 2 parts wheat
- 2 parts alfalfa meal or chopped hay
- 1 part meat scraps, fish meal, or soybean meal
- 2 to 3 parts dried split peas, lentils, or soybean meal
- 2 to 3 parts bone meal, crushed oyster shell, granite grit, or limestone
- ½ part cod-liver oil

You may also wish to add sunflower seeds, hulled barley, millet, kamut, amaranth seeds, quinoa, sesame seeds, flax seeds, or kelp granules. If you find that your eggs are thin-shelled, try adding more calcium to the feed (in the form of limestone or oystershell). Store feed in a covered bucket, barrel, or other container that will not allow rodents to get into it. A plastic or galvanized bucket is good, as it will also keep mold-causing moisture out of the feed.

the hen or the eggs. Make the nest of damp soil a few inches deep placed in the bottom of the box, and then lay sweet hay or clean straw on top of that.

Place the nesting box in a quiet and secluded place away from the other chickens. If space permits, you can construct a smaller shed in which to house your nesting hen. A hen can generally sit on anywhere between nine and fifteen eggs. The hen should only be allowed to leave the nest to feed, drink water, and take a dust bath. When the hen does leave her box, check the eggs and dispose of any damaged ones. An older hen will generally be more careful and apt to roost than a younger female.

Once the chicks are hatched, they will need to stay warm and clean, have lots of exercise, and have access to food regularly. Make sure the feed is ground finely enough that the chicks can easily eat and digest it. They should also have clean, fresh water.

BACTERIA ASSOCIATED WITH CHICKEN MEAT

- **Salmonella**—This is primarily found in the intestinal tract of poultry and can be found in raw meat and eggs.
- **Campylobacter jejuni**—This is one of the most common causes of diarrheal illness in humans, and is spread by improper handling of raw chicken meat and not cooking the meat thoroughly.
- **Listeria monocytogenes**—This causes illness in humans and can be destroyed by keeping the meat refrigerated and by cooking it thoroughly.

STORING EGGS

Eggs are among the most nutritious foods on earth and can be part of a healthy diet. Hens typically lay eggs every twenty-five hours, so you can be sure to have a fresh supply on a daily basis, in many cases. But eggs, like any other animal by-product, need to be handled safely and carefully to avoid rotting and spreading disease. Here are a few tips on how to best preserve your farm-fresh eggs:

1. Make sure your eggs come from hens that have not been running with roosters. Infertile eggs last longer than those that have been fertilized.
2. Keep the fresh eggs together.
3. Choose eggs that are perfectly clean.
4. Make sure not to crack the shells, as this will taint the taste and make the egg rot much quicker.
5. Place your eggs directly in the refrigerator where they will keep for several weeks.

Ducks

Ducks tend to be somewhat more difficult than chicks to raise, but they do provide wonderful eggs and meat. Ducks tend to have pleasanter personalities than chickens and are often prolific layers. The eggs taste similar to chicken eggs, but are usually larger and have a slightly richer flavor. Ducks are happiest and healthiest when they have access to a pool or pond to paddle around in and when they have several other ducks to keep them company.

Breeds of Ducks

There are six common breeds of ducks: White Pekin, White Aylesbury, Colored Rouen, Black Cayuga, Colored Muscovy, and White Muscovy. Each breed is unique and has its own advantages and disadvantages.

According to Mrs. Beeton in her *Book of Household Management*, published in 1861, "[Aylesbury ducks'] snowy plumage and comfortable comportment make it a credit to the poultry-yard, while its broad and deep breast, and its ample back, convey the assurance that your satisfaction will not cease at its death."

1. White Pekin—The most popular breed of duck, these are also the easiest to raise. These ducks are hardy and do well in close confinement. They are timid and must be handled carefully. Their large frame gives them lots of meat and they are also prolific egg layers.
2. White Aylesbury—This breed is similar to the Pekin but the plumage is much whiter and they are a bit heavier

White Pekins were originally bred from the Mallard in China and came to the United States in 1873.

than the former. They are not as popular in the United States as the White Pekin duck.

3. Colored Rouens—These darkly plumed ducks are also quite popular and fatten easily for meat purposes.
4. Black Cayuga and Muscovy breeds—These are American breeds that are easily raised but are not as productive as the White Pekin.

Housing Ducks

You don't need a lot of space in which to raise ducks—nor do you need water to raise them successfully, though they will be happier if you can provide at least a small pool of water for them to bathe and paddle around in. Housing for ducks is relatively simple. The houses do not have to be as warm or dry as for chickens but the ducks cannot be confined for long periods as chickens can. They

A Black Cayuga (bottom) stands with two Saxony ducks.

need more exercise out of doors in order to be healthy and to produce more eggs. A house that is protected from dampness or excess rain water and that has straw or hay covering the floor is adequate for ducks. If you want to keep your ducks somewhat confined, a small fence about 2½ feet high will do the trick. Ducks don't require nesting boxes, as they lay their eggs on the floor of the house or in the yard around the house.

Feeding and Watering Ducks

Ducks require plenty of fresh water to drink, as they have to drink regularly while eating. Ducks eat both vegetable and animal foods. If allowed to roam free and to find their own food stuff, ducks will eat grasses, small fish, and water insects (if streams or ponds are provided).

Ducks need their food to be soft and mushy in order for them to digest it. Ducklings should be fed equal parts corn meal, wheat bran, and flour for the first week of life. Then, for the next fifty days or so, the ducklings should be fed the above mixture in addition to a little grit or sand and some green foods (green rye, oats, clover) all mixed together. After this time, ducks should be fed on a mixture of 2 parts cornmeal, 1 part wheat bran, 1 part flour, some coarse sand, and green foods.

Hatching Ducklings

The natural process of incubation (hatching ducklings underneath a hen) is the preferred method of hatching ducklings. It is important to take good care of the setting hen. Feed her

whole corn mixed with green food, grit, and fresh water. Placing the feed and water just in front of the nest for the first few days will encourage the hen to eat and drink without leaving the nest. Hens will typically lay their eggs on the ground, in straw or hay that is provided for them. Make sure to clean the houses and pens often so the laying ducks have clean areas in which to incubate their eggs.

Caring for Ducklings

Young ducklings are very susceptible to atmospheric changes. They must be kept warm and free from getting chilled. The ducklings are most vulnerable during the first three weeks of life; after that time, they are more likely to thrive to adulthood. Construct brooders for the young ducklings and keep them very warm by hanging strips of cloth over the door cracks. After three weeks in the warm brooder, move the ducklings to a cold brooder as they can now withstand fluctuating temperatures.

Common Diseases

Ducks are not as prone to the typical poultry diseases, and many of the diseases they do contract can be prevented by making sure the ducks have a clean environment in which to live (by cleaning out their houses, providing fresh drinking water, and so on).

Two common diseases found in ducks are botulism and maggots. Botulism causes the duck's neck to go limp, making it difficult or even impossible for the duck to swallow. Maggots infest the ducks if they do not have any clean water in which to bathe, and are typically contracted in the hot summer months. Both of these diseases (as well as worms and mites) can be cured with the proper care, medications, and veterinary assistance.

Turkeys

Turkeys are generally raised for their meat (especially for holiday roasts) though their eggs can also be eaten. Turkeys are incredibly easy to manage and raise as they primarily subsist on bugs, grasshoppers, and wasted grain that they find while wandering around the yard. They are, in a sense, self-sustaining foragers.

If you are looking to raise a turkey for Thanksgiving dinner, it is best to hatch the turkey chick in early spring, so that by November, it will be about 14 to 20 pounds.

Breeds of Turkeys

The largest breeds of turkeys found in the United States are the Bronze and Narragansett. Other breeds, though not as popular, include the White Holland, Black turkey, Slate turkey, and Bourbon Red.

Bronze breeds are most likely a cross between a wild North American turkey and domestic turkey, and they have beautiful rich plumage. This is the most common type of turkey to raise, as it is the largest, is very hardy, and is the most profitable. The White Holland and Bourbon Red, however, are said to be the most "domesticated" in their habits and are easier to keep in a smaller roaming area.

Bronze turkeys like this one are some of the most common in the United States.

Housing Turkeys

Turkeys flourish when they can roost in the open. They thrive in the shelter of trees, though this can become problematic as they are more vulnerable to predators than if they are confined in a house. If you do build a house for them, it should be airy, roomy, and very clean.

It is important to allow turkeys freedom to roam; if you live in a more suburban or neighborhood area, raising turkeys may not be the best option for you as your turkeys may wander into a neighboring yard, upsetting your neighbors. Turkeys need lots of exercise to be healthy and vigorous. When turkeys are confined for large periods of time, it is more difficult to regulate their feeding (turkeys are natural foragers and thrive best on natural foods), and they are more likely to contract disease than if they are allowed to range freely.

What Do Turkeys Eat?

Turkeys gain most of their sustenance from foraging, either in lawns or in pastures. They typically eat green vegetation, berries, weed seeds, waste grain, nuts, and various kinds of acorns. In the summer months, turkeys especially like to get grasshoppers. Due to their love of eating insects that can damage crops and gardens, turkeys are quite useful in keeping your growing produce free from harmful insects and parasites.

Turkeys may be fed grain (similar to a mixture given to chickens) if they are going to be slaughtered, in order to make them larger.

Hatching Turkey Chicks

Turkey hens lay eggs in the middle of March to the first of April. If you are looking to hatch and raise turkey chicks, it is vital to watch the hen closely for when she lays the eggs, and then gather them and keep the eggs warm until the weather is more stable. Turkey hens generally aim to hide their nests from predators. It is best, for the hen's sake, to provide her with a coop of some sort, which she can freely enter and leave. Or, if no coop is available, encourage the hen to lay her eggs in a nest close to your house (putting a large barrel on its side and heaping up brush near the house may entice the hen

to nest there). This way, you can keep an eye on the eggs and hatchlings.

Hens are well adapted to hatch all of the eggs that they lay. It takes twenty-seven to twenty-nine days for turkey eggs to hatch. While the hens are incubating the eggs, they should be given adequate food and water, placed close to their nest. Wheat and corn are the best food during the laying and incubation period.

Raising the Poults

Turkey chicks, also known as "poults," can be difficult to raise and require lots of care and attention for their first few weeks of life. In this sense, a turkey raiser must be "on call" to come to the aid of the hen and her poults at any time during the day for the first month or so. Many times, the hens can raise the poults quite well, but it is important that they receive enough food and warmth in the early weeks to allow them to grow healthy and strong. The poults should stay dry, as they become chilled easily. If you are able, encouraging the poults and their mother into a coop until the poults are stronger will aid their growth to adulthood.

Poults should be fed soft and easily digestible foods. Stale bread, dipped in milk and then dried until it crumbles is an excellent source of food for the young turkeys.

Diseases

Turkeys are hardy birds but they are susceptible to a few debilitating or fatal diseases. It is a fact that the mortality rate among young turkeys, even if they are given all the care and exercise and food needed, is relatively high (usually due to environmental and predatory factors).

The most common disease in turkeys is blackhead. Blackhead typically infects young turkeys between six weeks and four months old. This disease will turn the head darker colored or even black and the bird will become very weak, will stop eating, and will have an insatiable thirst. Blackhead is usually fatal.

Another disease that turkeys occasionally contract is roup. Roup generally occurs when a turkey has been exposed to extreme dampness or cold drafts for long periods of time. Roup causes the turkey's head to swell around the eyes and is highly contagious to other turkeys. Nutritional roup is caused by a vitamin A deficiency, which can be alleviated by adding vitamin A to the turkeys' drinking water. It is best to consult a veterinarian if your turkey seems to have this disease.

SLAUGHTERING POULTRY

If you are raising your own poultry, you may decide that you'd like to use them for consumption as well. Slaughtering your own poultry enables you to know exactly what is in the meat you and your family are consuming, and to ensure that the poultry is kept humanely before being slaughtered. Here are some guidelines for slaughtering poultry:

1. To prepare a fowl for slaughter, make sure the bird is secured well so it is unable to move (either hanging down from a pole or laid on a block that is used for chopping wood).
2. Killing the fowl can be done in two ways: one way is to hang the bird upside down and to cut the jugular vein with a sharp knife. It is a good idea to have a funnel or vessel available to collect the draining blood so it does not make a mess and can be disposed of easily. The other option is to place the bird's head on a chopping block and then, in one clean movement, chop its head off at the middle of the neck. Then, hang the bird upside down and let the blood drain as described above.
3. Once the bird has been thoroughly drained of blood, you can begin to pluck it. Have a pot of hot water (around 140°F) ready, into which to dip the bird. Holding the bird by the feet, dip it into the pot of hot water and leave it for about forty-five seconds—you do not want the bird to begin to cook! Then, remove the bird from the pot and begin plucking immediately. The feathers should come off fairly easily, but this process takes time, so be patient. Discard the feathers.
4. Once the bird has been completely rid of feathers, slip back the skin from the neck and cut the neck off close to the base of the body. Then, remove the crop, trachea, and esophagus from the bird by loosening them and pulling them out through the hole created from chopping off the neck. Cut off the vent to release the main entrails (being careful not to puncture the intestines or bacteria could be released into the meat) and make a horizontal slit about an inch above it so you can insert two fingers. Remove the entrails, liver (carefully cutting off the gallbladder), gizzard, and heart from the bird and set the last three aside if you want to eat them later or make them into stuffing. If you are going to save the heart, slip off the membrane enclosing it and cut off the veins and arteries. Make sure to clean out the gizzard as well if you will be using it later.
5. Wash the bird thoroughly, inside and out, and wipe it dry.
6. Cut off the feet below the joints and then carefully pull out the tendons from the drumsticks.
7. Once the carcass is thoroughly dry and clean, store it in the refrigerator if it will be used that same day or the next. If you want to save the bird for later use, place it in a moisture-proof bag and set it in the freezer (along with any innards that you may have saved).
8. Make sure you clean and disinfect any surface you were working on to avoid the spread of bacteria and other diseases.

Rabbits

Rabbits are very social and docile animals, and easy to maintain. They like to play, but because of their skittish nature, are not necessarily the best pets for young children. Larger rabbits, bred for eating, often make good pets because of their more relaxed personalities. Rabbits are easier to raise than chickens and can provide you with beautiful fur and lean meat. In fact, rabbits will take up less space and use less money than chickens.

Breeds

There are over forty breeds of domestic rabbits. Below are ten of the most commonly owned varieties, along with their traits and popular uses.

1. Californian: 6–10 lbs. Short fur. Relaxed personality. Choice for eating.
2. Dutch: 3–5 lbs. Short fur. Relaxed personality. Choice pet. Good for young children.
3. Flemish Giant: 9+ lbs. Medium-length fur. Calm personality. Choice for eating.
4. Holland Lop: 3–5 lbs. Medium-length fur. Curious personality. Choice pet. One of the lop-eared rabbits, its ears flop down next to its face. A similar popular breed is the American Fuzzy Lop.

5. Jersey Wooly: 2–4 lbs. Long fur. Relaxed personality. Choice pet.
6. Mini Lop: 4–7 lbs. Medium-length fur. Relaxed personality. Choice pet. Lop-eared. Some reports of higher biting tendencies.
7. Mini Rex: 3–5 lbs. Very short, velvety fur. Curious personality. Choice pet. Tend to have sharp toenails.
8. Netherland Dwarf: 2–4 lbs. Medium-length fur. Excitable personality. Choice pet.
9. New Zealand: 9+ lbs. Short fur. Curious personality. Choice for eating. Variable reputation for biting.
10. Satin: 9+ lbs. Medium-length fur. Relaxed personality. Fur is finer and denser than other furs.

Housing

Rabbits should be kept in clean, dry, spacious homes. You will need a hutch, similar to a henhouse, to house your rabbits. It is important to provide your rabbits with lots of air. The best hutch will have a wide, overhanging roof and is elevated about six inches off the ground. This way, your rabbits will not only have shade, but their homes will be prevented from getting damp.

Food

A rabbit's diet should be made up of three things: a small portion pellets (provided they are high in fiber), a continual source of hay, and vegetables. Rabbits love vegetables that are dark and leafy or root vegetables. Avoid feeding them beans or rhubarb, and limit the amount of spinach they eat. If you want to give rabbits a treat, try a small piece of fruit, such as a banana or apple. Remember that all of their food needs to be fresh (pellets should not be more than six weeks old), and like all other animals, be careful not to overfeed them. Also, to keep them from dehydrating, provide them with plenty of clean water every day.

Note: If you have a pregnant doe, allow her to eat a little more than usual.

Breeding

When you want to breed rabbits, put a male and female together in the morning or evening. After they have mated, you may separate them again. A female's gestation period is approximately a month in length, and litters range from six to ten babies. Baby rabbits' eyes will not open until two weeks after birth. Their mother will nurse them for a month, and for at least the first week, you must not touch any of the litter; you can alter their smell and the mother may stop feeding them. At two months, babies should be weaned from their mother, and at four months, or approximately 4.5 pounds, they are old enough to sell, eat, or continue breeding. Larger rabbit varieties may take six to twelve months to sexually mature.

Health Concerns

The main issues that may arise in your rabbits' health are digestive problems and bacterial infections. Monitor your rabbit's droppings carefully. Diarrhea in rabbits can be fatal. Some diarrhea is easy to identify, but also be on the lookout for droppings that are misshapen, softer in consistency, a lack of droppings altogether, and loud tummy growling. Diarrhea requires antibiotics from your veterinarian. In bacterial infections, your rabbit may have a runny nose or eyes, a high temperature, or a rattling or coughing respiratory noise. This also requires medical attention and an antibiotic specific to the type of infection.

Hairballs are another issue you may encounter and also require some attention. Every three months, rabbits shed their hair, and these sheds will alter between light and heavy. Since rabbits will attempt to groom themselves as cats do, but cannot vomit hair as cats can, you must groom them additionally, to prevent too much hair ingestion. Brush and comb them when their shedding begins, and provide them with ample fresh hay and opportunity for exercise. The fiber in the hay will help the hair to pass through their digestive tracts, and the exercise will keep their metabolisms active.

If your rabbit has badly misaligned teeth, they may interfere with his or her ability to eat and will need to be trimmed by the veterinarian.

Never give your rabbit amoxicillin or use cedar or pine shavings in their hutches. Penicillin-based drugs carry high risks for rabbits, and the shavings emit a carbon that can cause respiratory or liver damage to small animals like rabbits.

Beekeeping

Beekeeping (also known as apiculture) is one of the oldest human industries. For thousands of years, honey has been considered a highly desirable food. Beekeeping is a science and can be a very profitable employment; it is also a wonderful hobby for many people in the United States. Keeping bees can be done almost anywhere—on a farm, in a rural or suburban area, and even, at times, in urban areas (even on rooftops!). Anywhere there are sufficient flowers from which to collect nectar, bees can thrive.

Apiculture relies heavily on the natural resources of a particular location and the knowledge of the beekeeper in order to be successful. Collecting and selling honey at your local farmers' market or just to family and friends can supply you with some extra cash if you are looking to make a profit from your apiary.

Why Raise Bees?

Bees are essential in the pollination and fertilization of many fruit and seed crops. If you have a garden with many flowers or fruit plants, having bees nearby will only help your garden flourish and grow year after year. Furthermore, nothing is more satisfying than extracting your own honey for everyday use.

How to Avoid Getting Stung

Though it takes some skill, you can learn how to avoid being stung by the bees you keep. Here are some ways you can keep your bee stings to a minimum:

1. Keep gentle bees. Having bees that, by sheer nature, are not as aggressive will reduce the number of stings you are likely to receive. Carniolan bees are one of the gentlest species, and so are the Caucasian bees introduced from Russia.
2. Obtain a good "smoker" and use it whenever you'll be handling your bees. Pumping smoke of any kind into and around the beehive will render your bees less aggressive and less likely to sting you.
3. Purchase and wear a veil. This should be made out of black bobbinet and worn over your face. Also, rubber gloves help protect your hands from stings.
4. Use a "bee escape." This device is fitted into a slot made in a board the same size as the top of the hive. Slip the board into the hive before you open it to extract the honey, and it allows the worker bees to slip below it but not to return back up. So, by placing the "bee escape" into the hive the day before you want to gain access to the combs and honey, you will most likely trap all the bees under the board and leave you free to work with the honeycombs without fear of stings.

What Type of Hive Should I Build?

Most beekeepers would agree that the best hives have suspended, moveable frames where the bees make the honeycombs, which are easy to lift out. These frames, called Langstroth frames, are the most popular kind of frame used by apiculturists in the United States.

Whether you build your own beehive or purchase one, it should be built strongly and should contain accurate bee spaces and a close-fitting, rainproof roof. If you are looking to have honeycombs, you must have a hive that permits the insertion of up to eight combs.

Where Should the Hive Be Situated?

Hives and their stands should be placed in an enclosure where the bees will not be disturbed by other animals or humans and where it will be generally quiet. Hives should be placed on their own stands at least 3 feet from each other. Do not allow weeds to grow near the hives and keep the hives away from walls and fences. You, as the beekeeper, want to be able to easily access your hive without fear of obstacles.

Swarming

Swarming is simply the migration of honeybees to a new hive and is led by the queen bee. During swarming season (the warm summer days), a beekeeper must remain very alert. If you see swarming above the hive, take great care and act calmly and quietly. You want to get the swarm into your hive, but this will be tricky. It they land on a nearby branch or in a basket, simply approach and then "pour" them into the hive. Keep in mind

that bees will more likely inhabit a cool, shaded hive than one that is baking in the hot summer sun.

Sometimes it is beneficial to try to prevent swarming, such as if you already have completely full hives. Removing the new honey frequently from the hive before swarming begins will deter the bees from swarming. Shading the hives on warm days will also help keep the bees from swarming.

Bee Pastures

Bees will fly a great distance to gather food but you should try to contain them, as well as possible, to an area within 2 miles of the beehive. Make sure they have access to many honey-producing plants, which you can grow in your garden. Alfalfa, asparagus, buckwheat, chestnut, clover, catnip, mustard, raspberry, roses, and sunflowers are some of the best honey-producing plants and trees. Also, make sure that your bees always have access to pure, clean water.

Preparing Your Bees for Winter

If you live in a colder region of the United States, keeping your bees alive throughout the winter months is difficult. If your queen bee happens to die in the fall, before a young queen can be reared, your whole colony will die throughout the winter. However, the queen's death can be avoided by taking simple precautions and giving careful attention to your hive come autumn.

Colonies are usually lost in the winter months due to insufficient winter food storages, faulty hive construction, lack of protection from the cold and dampness, not enough or too much ventilation, or too many older bees and not enough young ones.

If you live in a region that gets a few weeks of severe weather, you may want to move your colony indoors, or at least to an area that is protected from the outside elements. But the essential components of having a colony survive through the winter season are to have a good queen; a fair ratio of healthy, young, and old bees; and a plentiful supply of food. The hive needs to retain a liberal supply of ripened honey and a thick syrup made from white cane sugar (you should feed this to your bees early enough so they have time to take the syrup and seal it over before winter).

To make this syrup, dissolve 3 pounds of granulated sugar in 1 quart of boiling water and add 1 pound of pure extracted honey to this. If you live in an extremely cold area, you may need up to 30 pounds of this syrup, depending on how many bees and hives you have. You can either use a top feeder or a frame feeder, which fits inside the hive in the place of a frame. Fill the frame with the syrup and place sticks or grass in it to keep the bees from drowning.

Extracting Honey

To obtain the extracted honey, you'll need to keep the honeycombs in one area of the hive or packed one above the other. Before removing the filled combs, you should allow the bees ample time to ripen and cap the

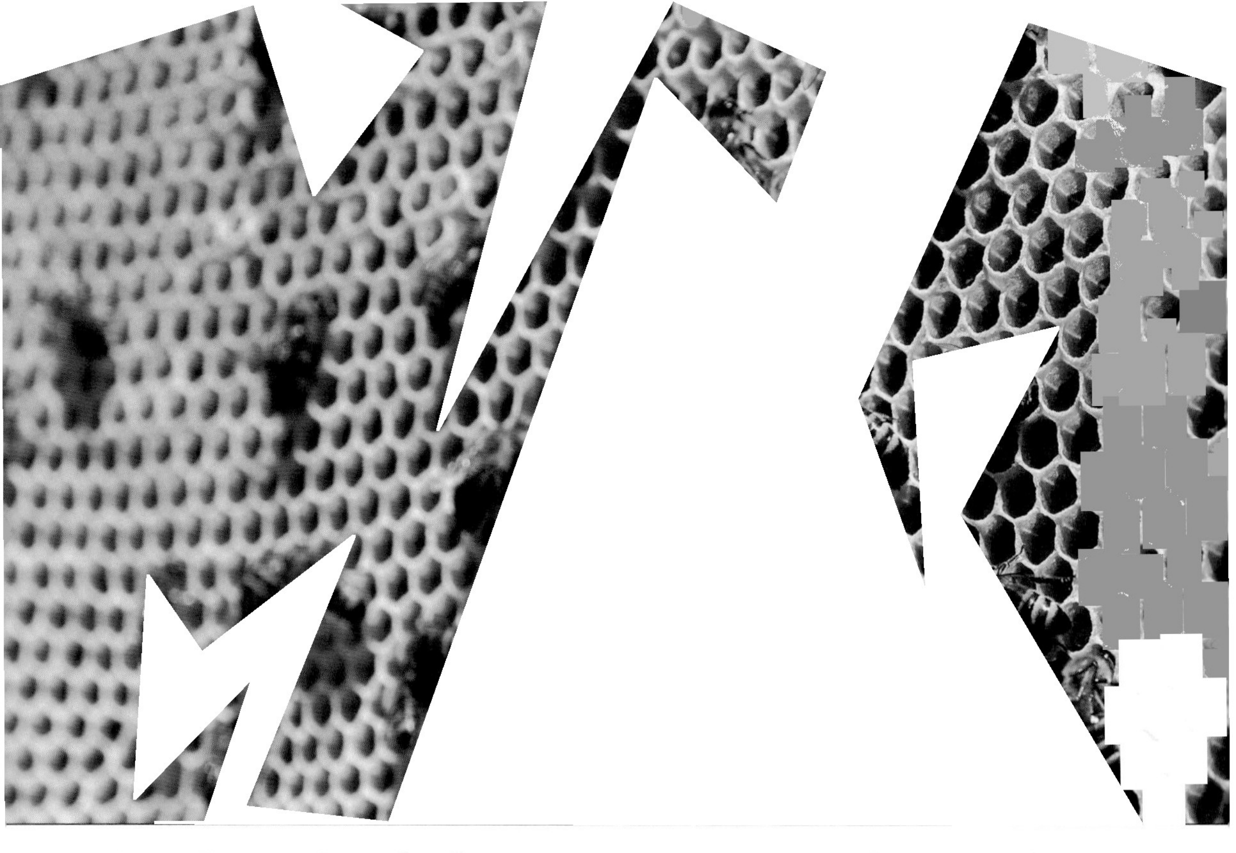

honey. To uncap the comb cells, simply use a sharp knife (apiary suppliers sell knives specifically for this purpose). Then put the combs in a machine called a honey extractor to extract the honey. The honey extractor whips the honey out of the cells and allows you to replace the fairly undamaged comb into the hive to be repaired and refilled.

The extracted honey runs into open buckets or vats and is left, covered with a tea towel or larger cloth, to stand for a week. It should be in a warm, dry room where no ants can reach it. Skim the honey each day until it is perfectly clear. Then you can put it into cans, jars, or bottles for selling or for your own personal use.

Making Beeswax

Beeswax from the honeycomb can be used for making candles, can be added to lotions or lip balm, and can even be used in baking. Rendering wax in boiling water is especially simple when you only have a small apiary.

Collect the combs, break them into chunks, roll them into balls if you like, and put them in a muslin bag. Put the bag with the beeswax into a large stockpot and bring the water to a slow boil, making sure the bag doesn't rest on the bottom of the pot and burn. The muslin will act as a strainer for the wax. Use clean, sterilized tongs to occasionally squeeze the bag. After the wax is boiled out of the bag, remove the pot from the heat and allow it to cool. Then, remove the wax from the top of the water and then remelt it in another pot on very low heat so it doesn't burn.

Pour the melted wax into molds lined with wax paper or plastic wrap and then cool it before using it to make other items or selling it at your local farmers' market.

Extra Beekeeping Tips

General Tips

1. Clip the old queen's wings and go through the hives every ten days to destroy queen cells to prevent swarming.
2. Always act and move calmly and quietly when handling bees.
3. Keep the hives cool and shaded. Bees won't enter a hot hive.

When Opening the Hive

1. Have a smoker ready to use if you desire.
2. Do not stand in front of the hive while the bees are entering and exiting.
3. Do not drop any tools into the hive while it's open.
4. Do not run if you become frightened.
5. If you are attacked, move away slowly and smoke the bees off yourself as you retreat.
6. Apply ammonia or a paste of baking soda and water immediately to any bee sting to relieve the pain. You can also scrape the area of the bee sting with your fingernail or the dull edge of a knife immediately after the sting.

When Feeding Your Bees

1. Keep a close watch over your bees during the entire season to see if they are feeding well or not.
2. Feed the bees during the evening.
3. Make sure the bees have ample water near their hive, especially in the spring.

Making a Beehive

The most important parts of constructing a beehive are to make it simple and sturdy. Just a plain box with a few frames and a couple of other loose parts will make a successful beehive that will be easy to use and manipulate. It is crucial that your beehive be well adapted to the nature of bees and also the climate where you live. Framed hives usually suffice for the beginning beekeeper. Below is a diagram of a simple beehive that you can easily construct for your backyard beekeeping purposes.

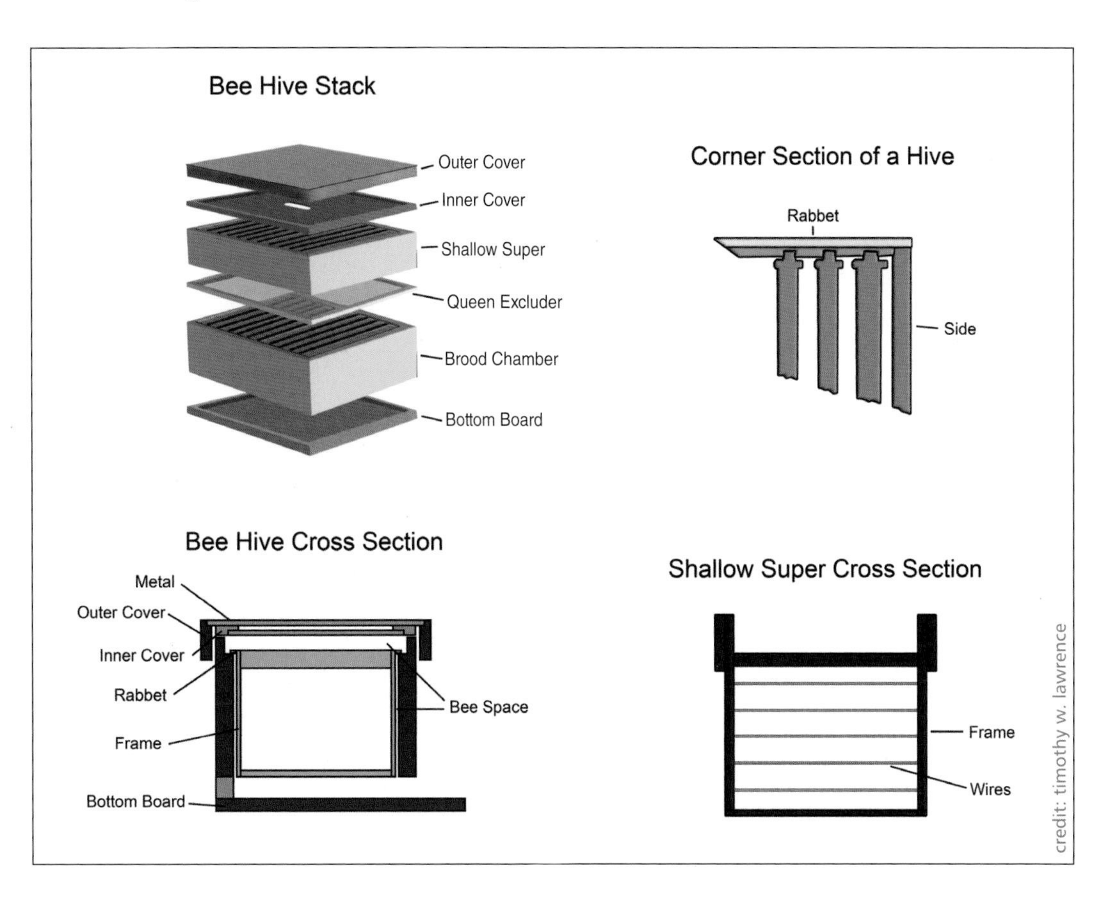

Goats

Goats provide us with milk and wool and thrive in arid, semitropical, and mountainous environments. In the more temperate regions of the world, goats are raised as supplementary animals, providing milk and cheese for families and acting as natural weed killers.

Breeds of Goats

There are many different types of goats. Some breeds are quite small (weighing roughly 20 pounds) and some are very large (weighing up to 250 pounds). Depending on the breed, goats may have horns that are corkscrew in shape, though many domestic goats are dehorned early on to lessen any potential injuries to humans or other goats. The hair of goats can also differ—various breeds have short hair, long hair, curly hair, silky hair, or coarse hair. Goats come in a variety of colors (solid black, white, brown, or spotted).

Feeding Goats

Goats can sustain themselves on bushes, trees, shrubs, woody plants, weeds, briars, and herbs. Pasture is the lowest cost feed available for goats, and allowing goats to graze

SIX MAJOR U.S. GOAT BREEDS

Alpine—Originally from Switzerland, these goats may have horns, are short-haired, and are usually white and black in color. They are also good producers of milk.

Anglo-Nubian—A cross between native English goats and Indian and Nubian breeds, these goats have droopy ears, spiral horns, and short hair. They are quite tall and do best in warmer climates. They do not produce as much milk, though it is much higher in fat than other goats. They are the most popular breed of goat in the United States.

LaMancha—A cross between Spanish Murciana and Swiss and Nubian breeds, these goats are extremely adaptable, have straight noses, short hair, may have horns, and do not have external ears. They are not as good milk producers as the Saanen and Toggenburg breeds, and their milk fat content is much higher.

Pygmy—Originally from Africa and the Caribbean, these dwarfed goats thrive in hotter climates. For their size, they are relatively good producers of milk.

Saanen—Originally from Switzerland, these goats are completely white, have short hair, and sometimes have horns. Goats of this breed are wonderful milk producers.

Toggenburg—Originally from Switzerland, these goats are brown with white facial, ear, and leg stripes; have straight noses; may have horns; and have short hair. This breed is very popular in the United States. These goats are good milk producers in the summer and winter seasons and survive well in both temperate and tropical climates.

in the summer months is a wonderful and economical way to keep goats, even if your yard is quite small. Goats thrive best when eating alfalfa or a mixture of clover and timothy. If you have a lawn and a few goats, you don't need a lawn mower if you plant these types of plants for your goats to eat. The one drawback to this is that your goats (depending on how many you own) may quickly deplete these natural resources, which can cause weed growth and erosion. Supplementing pasture feed with other food stuff, such as greenchop, root crops, and wet brewery grains, will ensure that your yard does not become overgrazed and that your goats remain well-fed and healthy. It is also beneficial to supply your goats with unlimited access to hay while they are grazing. Make sure that your goats have easy access to shaded areas and fresh water, and offer a salt and mineral mix on occasion.

Dry forage is another good source of feed for your goats. It is relatively inexpensive to grow or buy and consists of good quality legume hay (alfalfa or clover). Legume hay is high in protein and has many essential minerals beneficial to your goats. To make sure your forages are highly nutritious, be sure that there are many leaves that provide protein and minerals and that the forage had an early cutting date, which will allow for easier digestion of the nutrients. If your forage is green in color, it most likely contains more vitamin A, which is good for promoting goat health.

Goat Milk

Goat milk is a wonderful substitute for those who are unable to tolerate cow's milk,

or for the elderly, babies, and those suffering from stomach ulcers. Milk from goats is also high in vitamin A and niacin but does not have the same amount of vitamins B6, B12, and C as cow's milk.

Lactating goats do need to be fed the best quality legume hay or green forage possible, as well as grain. Give the grain to the doe at a rate that equals ½ pound of grain for every pound of milk she produces.

Common Diseases Affecting Goats

Goats tend to get more internal parasites than other herd animals. Some goats develop infectious arthritis, pneumonia, coccidiosis, scabies, liver fluke disease, and mastitis. It is advisable that you establish a relationship with a good veterinarian who specializes in small farm animals to periodically check your goats for various diseases.

Milking a Goat

Milking a goat takes some practice and patience, especially when you first begin. However, once you establish a routine and rhythm to the milking, the whole process should run relatively smoothly. The main thing to remember is to keep calm and never pull on the teat, as this will hurt the goat and she might upset the milk bucket. The goat will pick up on any anxiousness or nervousness on your part and it could affect how cooperative she is during the milking.

Supplies

- A grain bucket and grain for feeding the goat while milking is taking place
- Milking stand
- Metal bucket to collect the milk
- A stool to sit on (optional)
- A warm sterilized wipe or cloth that has been boiled in water
- Teat dip solution (2 tbsps bleach, 1 quart water, one drop normal dish detergent mixed together)

Directions

1. Ready your milking stand by filling the grain bucket with enough grain to last throughout the entire milking. Then retrieve the goat, separating her from any other goats to avoid distractions and unsuccessful milking. Place the goat's head through the head hold of the milking stand so she can eat the grain and then close the lever so she cannot remove her head.
2. With the warm, sterilized wipe or cloth, clean the udder and teats to remove any dirt, manure, or bacteria that may be present. Then, place the metal bucket on the stand below the udder.
3. Wrap your thumb and forefinger around the base of one teat. This will help trap the milk in the teat so it can be squirted out. Then, starting with your middle finger, squeeze the three remaining fingers in one single, smooth motion to squirt the milk into the bucket. Be sure to keep a tight grip on the base of the teat so the milk stays there until extracted. Remember: the first squirt of milk from either teat should not be put into the bucket as it may contain dirt or bacteria that you don't want contaminating the milk.
4. Release the grip on the teat and allow it to refill with milk. While this is happening, you can repeat this process on the other teat and can alternate between teats to speed up the milking process.
6. When the teats begin to look empty (they will be somewhat flat in appearance), massage the udder just a little bit to see if any more milk remains. If so, squeeze it out in the same manner as above until you cannot extract much more.
7. Remove the milk bucket from the stand and then, with your teat dip mixture in a disposable cup, dip each teat into the solution and allow to air dry. This will

keep bacteria and infection from going into the teat and udder.

8. Remove the goat from the milk stand and return her to the pen.

Making Cheese from Goat Milk

Most varieties of cheese that can be made from cow's milk can also be successfully made using goats' milk. Goats' milk cheese can easily be made at home. In order to make the cheese, however, at least one gallon of goat milk should be available. Make sure that all of your equipment is washed and sterilized (using heat is fine) before using it.

Cottage Cheese

1. Collect surplus milk that is free of strong odors. Cool it to around 40°F and keep it at that temperature until it is used.
2. Skim off any cream. Use the skim milk for cheese and the cream for cheese dressing.
3. If you wish to pasteurize your milk (which will allow it hold better as a cheese), collect all the milk to be processed into a flat bottomed, straight-sided pan and heat to 145°F on low heat. Hold it at this temperature for about thirty minutes and then cool to around 80°F. Use a dairy thermometer to measure the milk's temperature. Then, inoculate the cheese milk with a desirable lactic acid fermenting bacterial culture (you can use commercial buttermilk for the initial source). Add about 7 ounces to 1 gallon of cheese milk, stir well, and let it sit undisturbed for about ten to sixteen hours, until a firm curd is formed.
4. When the curd is firm enough, cut the curd into uniform cubes no larger than ½ inch using a knife or spatula.
5. Allow the curd to sit undisturbed for a couple of minutes and then warm it slowly, stirring carefully, at a temperature no greater than 135°F. The curd should eventually become firm and free from whey.
6. When the curd is firm, remove from the heat and stop stirring. Siphon off the excess whey from the top of the pot. The curd should settle to the bottom of the container. If the curd is floating, bacteria that produces gas has been released and a new batch must be made.
7. Replace the whey with cold water, washing the curd and then draining the water. Wash again with ice-cold water from the refrigerator to chill the curd. This will keep the flavor fresh.
8. Using a draining board, drain the excess water from the curd. Now your curd is complete.
9. In order to make the curd into a cottage cheese consistency, separate the curd as much as possible and mix with a milk or cream mixture containing salt to taste.

Domiati Cheese

This type of cheese is made throughout the Mediterranean region. It is eaten fresh or aged two to three months before consumption.

1. Cool a gallon of fresh, quality milk to around 105°F, adding 8 ounces of salt to the milk. Stir the salt until it is completely dissolved.
2. Pasteurize the milk as described in step 3 of the cottage cheese recipe above.
3. Domiati cheese is coagulated by adding a protease enzyme (rennet). This enzyme may be purchased at a local drug store, health food store, or a cheese maker in your area. Dissolve the concentrate in water, add it to the cheese milk, and stir for a few minutes. Use 1 milliliter of diluted rennet liquid in 40 milliliters of water for every 2½ gallons of cheese milk.
4. Set the milk at around 105°F. When the enzyme is completely dispersed in the cheese milk, allow the mix to sit undisturbed until it forms a firm curd.

ANGORA GOATS

Angora goats may be the most efficient fiber producers in the world. The hair of these goats is made into mohair, a long, lustrous hair that is woven into fine garments. Angora goats are native to Turkey and were imported to the United States in the mid-1800s. Now, the United States is one of the two biggest producers of mohair on Earth.

Angora goats are typically relaxed and docile. They are delicate creatures, easily strained by their year-round fleeces. Angora goats need extra attention and are more high-maintenance than other breeds of goat. While these goats can adapt to many temperate climates, they do particularly well in the arid environment of the southwestern states.

Angora goats can be sheared twice yearly, before breeding and before birthing. The hair of the goat will grow about ¾ inch per month and it should be sheared once it reaches 4 to 6 inches in length. During the shearing process, the goat is usually lying down on a clean floor with its legs tied. When the fleece is gathered (it should be sheared in one full piece), it should be bundled into a burlap bag and should be free of contaminants. Mark your name on the bag and make sure there is only one bag per fleece. For more thorough rules and regulations about selling mohair through the government's direct-payment program, contact the USDA Agricultural Stabilization and Conservation Service online or in one of their many offices.

Shearing can be accomplished with the use of a special goat comb, which leaves ¼ inch of stubble on the goat. It is important to keep the fleeces clean and to avoid injuring the animal. The shearing seasons are in the spring and fall. After a goat has been sheared, it will be more sensitive to changes in the weather for up to six weeks. Make sure you have proper warming huts for these goats in the winter and adequate shelter from rain and inclement weather.

5. When the desired firmness is reached, cut the curd into very small cubes. Allow for some whey separation. After ten to twenty minutes, remove and reserve about a third of the volume of salted whey.
6. Put the curd and remaining whey into cloth-lined molds (the best are rectangular stainless steel containers with perforated sides and bottom) with a cover. The molds should be between 7 and 10 inches in height. Fill the molds with the curd, fold the cloth over the top, allow the whey to drain, and discard the whey.
7. Once the curd is firm enough, apply added weight for ten to eighteen hours until it is as moist as you want.
8. Once the pressing is complete and the cheese is formed into a block, remove the molds, and cut the blocks into 4-inch-thick pieces. Place the pieces in plastic containers with airtight seals. Fill the containers with reserved salted whey from step 5, covering the cheese by about an inch.
9. Place these containers at a temperature between 60 and 65°F to cure for one to four months.

Feta Cheese

This type of cheese is very popular to make from goats' milk. The same process is used as the Domiati cheese except that salt is not added to the milk before coagulation. Feta cheese is aged in a brine solution after the cubes have been salted in a brine solution for at least twenty-four hours.

Sheep

Sheep were possibly the first domesticated animals, and are now found all over the world on farms and smaller plots of land. Almost all the breeds of sheep that are found in the United States have been brought here from Great Britain. Raising sheep is relatively easy, as they only need pasture to eat, shelter from bad weather, and protection from predators. Sheep's wool can be used to make yarn or other articles of clothing and their milk can be made into various types of cheeses and yogurt, though this is not normally done in the United States.

Sheep are naturally shy creatures and are extremely docile. If they are treated well, they will learn to be affectionate with their owner. If a sheep is comfortable with its owner, it will be much easier to manage and to corral into its pen if it's allowed to graze freely. Start with only one or two sheep; they are not difficult to manage but do require a lot of attention.

Breeds of Sheep

There are many different breeds of sheep—some are used exclusively for their meat and others for their wool. Six quality wool-producing breeds are as follows:

1. Cotswold Sheep—This breed is very docile and hardy and thrives well in pastures. It produces around 14 pounds of

fleece per year, making it a very profitable breed for anyone wanting to sell wool.
2. Leicester sheep—This is a hardy, docile breed of sheep that is a very good grazer. This breed has 6-inch-long, coarse wool that is desirable for knitting. It is a very popular breed in the United States.
3. Merino sheep—Introduced to the United States in the early twentieth century, this small- to medium-sized sheep has lots of rolls and folds of fine white wool and produces a fleece anywhere between 10 and 20 pounds. It is considered a fine-wool specialist, and though its fleece appears dark in color, the wool is actually white or buff. It is a wonderful foraging sheep, is hardy, and has a gentle disposition, but is not a very good milk producer.
4. Oxford Down sheep—A more recent breed, these dark-faced sheep have hardy constitutions and good fleece.
5. Shropshire sheep—This breed has longer, more open, and coarser fleece than other breeds. It is quite popular in the United States, especially in areas that are more moist and damp, as they seem to fare better in these climates than other breeds of sheep.
6. Southdown sheep—One of the oldest breeds of sheep, they are popular for their good quality wool and are deemed the standard of excellence for many sheep owners. Docile, hardy, and good grazing on pastures, their coarse and light-colored wool is used to make flannel.

Housing Sheep

Sheep do not require much shelter—only a small shed that is open on one side (preferably to the south so it can stay warmer in the winter months) and is roughly 6 to 8 feet high. The shelter should be ventilated well to reduce any unpleasant smells and to keep the sheep cool in the summer. Feeding racks or mangers should be placed inside of the shed to hold the feed for the sheep. If you live in a colder region of the country, building a sturdier, warmer shed for the sheep to live in during the winter is recommended.

Straw should be used for the sheep's bedding and should be changed daily to make sure the sheep do not become ill from an unclean shelter. Especially for the winter months, a dry pen should be erected for the sheep to exercise in. The fences should be strong enough to keep out predators that may enter your yard and to keep the sheep from escaping.

What Do Sheep Eat?

Sheep generally eat grass and are wonderful grazers. They utilize rough and scanty pasturage better than other grazing animals and, due to this, they can actually be quite beneficial in cleaning up a yard that is overgrown with undesirable herbage. Allowing sheep to graze in your yard or in a small pasture field will provide them with sufficient food in the summer months. Sheep also eat a variety of weeds, briars, and shrubs. Fresh water should always be available for the sheep every time of year.

During the winter months especially, when grass is scarce, sheep should be fed on hay (alfalfa, legume, or clover hay) and small quantities of grain. Corn is also a good win-

ter food for the sheep (it can also be mixed with wheat bran), and straw, salt, and roots can also be occasionally added to their diet. Good food during the winter season will help the sheep grow a healthier and thicker wool coat.

Shearing Sheep

Sheep are generally sheared in the spring or early summer before the weather gets too warm. To do your own shearing, invest in a quality hand shearer and a scale on which to weigh the fleece. An experienced shearer should be able to take the entire wool off in one piece.

You may want to wash the wool a few days to a week before shearing the sheep. To do so, corral the sheep into a pen on a warm spring day (make sure there isn't a cold breeze blowing and that there is a lot of sunshine so the sheep does not become chilled). Douse the sheep in warm water, scrub the wool, and rinse. Repeat this a few times until most of the dirt and debris is out of the wool. Diffuse some natural oil throughout the wool to make it softer and ready for shearing.

The sheep should be completely dry before shearing and you should choose a warm—but not overly hot—day. If you are a beginner at shearing sheep, try to find an experienced sheep owner to show you how to properly hold and shear a sheep. This way, you won't cause undue harm to the sheep's skin and will get the best fleece possible. When you are hand-shearing a sheep, remember to keep the skin pulled taut on the part where you are shearing to decrease the potential of cutting the skin.

Once the wool is sheared, tag it and roll it up by itself, and then bind it with twine. Be sure not to fold it or bind it too tightly. Separate and remove any dirty or soiled parts of the fleece before binding, as these parts will not be able to be carded and used.

Carding and Spinning Wool

To make the sheared wool into yarn you will need only a few tools: a spinning wheel or drop spindle and wool cards. Wool cards are rectangular pieces of thin board that have many wire teeth attached to them (they look like coarse brushes that are sometimes used for dogs' hair). To begin, you must clean the wool fleece of any debris, feltings, or other imperfections before carding it; otherwise your yarn will not spin correctly. Also wash it to remove any additional sand or dirt embedded in the wool and then allow it to dry completely. Then, all you need is to gather your supplies and follow some simple instructions.

Carding Wool

1. Grease the wool with rape oil or olive oil, just enough to work into the fibers.
2. Take one wool card in your left hand, rest it on your knee, gather a tuft of wool from the fleece, and place it onto the wool-card so it is caught between the wired teeth of the card.
3. Take the second wool card in your right hand and bring it gently across the other card several times, making a brushing movement toward your body.
4. When the fibers are all brushed in the same direction and the wool is soft and fluffy to the touch, remove the wool by rolling it into a small fleecy ball (roughly a foot or more in length and only 2 inches in width) and put it in a bag until it is used for spinning.

Note: Carded wool can also be used for felting, in which case no spinning is needed. To felt a small blanket, place large amounts of carded wool on either side of a burlap sack. Using felting needles, weave the wool into the burlap until it is tightly held by the jute or hemp fabrics of the burlap.

Spinning Wool

1. Take one long roll of carded wool and wind the fibers around the spindle.
2. Move the wheel gently and hold the spindle to allow the wool to "draw," or start to pull together into a single thread.
3. Keep moving the wheel and allow the yarn to wind around the spindle or a separate spool, if you have a more complex spinning wheel.
4. Keep adding rolls of carded wool to the spindle until you have the desired amount of yarn.

Note: If you are unable to obtain a spinning wheel of any kind, you can spin your carded wool by hand, although this will not produce the same tightness in your yarn as regular spinning. All you need to do is take the carded wool, hold it with one hand, and pull and twist the fibers into one, continuous piece. Winding the end of the yarn around a stick, spindle, or spool and securing it in place at the end will help keep your fibers tight and your yarn twisted.

If you want your yarn to be different colors, try dying it with natural berry juices or with special wool dyes found in arts and crafts stores.

Milking Sheep

Sheep's milk is not typically used in the United States for drinking, making cheese, or other familiar dairy products. Sheep do not typically produce milk year-round, as cows do, so milk will only be produced if you bred your sheep and had a lamb produced. If you do have a sheep that has given birth and the lamb has been sold or taken away, it is important to know how to milk her so her udders do not become caked. Some ewes will still have an abundance of milk even after their lambs have been weaned and this excess milk should be removed to keep the ewe healthy and her udder free from infection.

To milk a ewe, secure her to a sturdy pole or hook with a short lead. Wash the udders gently to remove any contaminants. Place the milk bucket below the udders and squeeze the teats downward, rhythmically, until the milk begins to flow into the bucket. Allowing the ewe to eat from a feed bucket while being milked will help to keep her relaxed. Strain and refrigerate the milk immediately.

Diseases

The main diseases to which sheep are susceptible are foot rot and scabs. These are contagious and both require proper treatment. Sheep may also acquire stomach worms if they eat hay that has gotten too damp or has been lying on the floor of their shelter. As always, it is best to establish a relationship with a veterinarian who is familiar with caring for sheep and have your flock regularly checked for any parasites or diseases that may arise.

Llamas

Llamas often make excellent pets and are a great source of wooly fiber (their wool can be spun into yarn). Llamas are being kept more and more by people in the United States as companion animals, sources of fiber, pack and light plow animals, therapy animals for the elderly, "guards" for other backyard animals, and good educational tools for children. Llamas have an even temperament and are very intelligent. Their intelligence and gentle nature make them easy to train, and their hardiness allows them to thrive well in both cold and warmer climates (although they can have heat stress in extremely hot and humid parts of the country).

Before you decide to purchase a llama or two for your yard, check your state requirements regarding livestock. In some places, your property must also be zoned for livestock.

Llamas come in many different colors and sizes. The average adult llama is between 5½ and 6 feet tall and weighs between 250 and 450 pounds. Llamas, being herd animals, like the company of other llamas, so it is advisable that you raise a pair to keep each other company. If you only want to care for one llama, then it would be best to also have a sheep, goat, or other animal that can be penned with the llama for camaraderie. Although llamas can be led well on a harness and lead, never tie one up as it could potentially break its own neck trying to break free.

Llamas tend to make their own communal dung heap in a particular part of their pen. This is quite convenient for cleanup

and allows you to collect the manure, compost it, and use it as a fertilizer for your garden.

Feeding Llamas

Llamas can subsist fairly well on grass, hay (an adult male will eat about one bale per week), shrubs, and trees, much like sheep and goats. If they are not receiving enough nutrients, they may be fed a mixture of rolled corn, oats, and barley, especially during the winter season when grazing is not necessarily available. Make sure not to overfeed your llamas, though, or they will become overweight and constipated. You can occasionally give cornstalks to your llamas as an added source of fiber, and you may add mineral supplements to the feed mixture or hay if you want. Salt blocks are also acceptable to have in your llama pen, and a constant supply of fresh water is necessary. Nursing female llamas should receive a grain mixture until the cria (baby) is weaned.

Be sure to keep feed and hay off the ground. This will help ward off parasites that establish themselves in the feed and are then ingested by the llamas.

Housing Your Llamas

Llamas may be sheltered in a small stable or even a converted garage. There should be enough room to store feed and hay, and the shelter should be able to be closed off during wet, windy, and cold weather. Llamas prefer light, open spaces in which to live, so make sure your shed or shelter has large doors and/or big windows. The feeders for the hay and grain mixture should be raised above the ground. Adding a place where a llama can be safely restrained for toenail clippings and vet checkups will help facilitate these processes but is not absolutely necessary.

The llamas should be able to enter and exit the shelter easily and it is a good idea to build a fence or pen around the shelter so they do not wander off. A fence about four feet tall should be enough to keep your llamas safe and enclosed. If you happen to have both a male and female llama, it is necessary to have separate enclosures for them to stave off unwanted pregnancies.

Toenail Trimming

Llamas need their toenails to be trimmed so they do not twist and fold under the toe, making it difficult for the llama to move around. Laying gravel in the area where your llamas frequently walk will help to keep the toenails naturally trimmed, but if you need to cut them, be careful not to cut too deeply or you may cause the tip of the toe to bleed and this could lead to an infection in the toe. Use shears designed for this purpose to cut the nails. Use one hand to hold the llama's "ankle" just above where the foot bends. Hold the clippers in your other hand, cutting away from the foot toward the tip of the nail. The nail's are easiest to clip in the early morning or after a rain, since the wetness of the ground will soften them.

Shearing

It is important to groom and shear your llama, especially during hot weather. Brushing the llama's coat to remove dirt and keep it from matting will not only make your llamas look clean and healthy but it will improve the quality of their coats. If you want to save the fibers for spinning into yarn, it is best to brush, comb, and use a hair dryer to remove any dust and debris from the llama's coat before you begin shearing.

Shearing is not necessarily difficult, but if you are a first-time llama owner, you should ask another llama farmer to teach you how to properly shear your llama. In order to shear your llama, you can purchase battery-operated shears to remove the fibers for sale or use. Different llamas will respond in different ways to shearing. Try holding the llama with a halter and lead in a smaller area to begin the shearing process. Do not completely remove the llama from any other llamas you have, though, as their presence will help calm the llama you are shearing. It is best to have another person with you to aid in the shearing (to hold the llama, give it treats, and offer any other help). When shearing a llama, don't shear all the way down to the skin. Allowing a thin coating of hair to cover the llama's body will help protect it from the sun and from being scratched when it rolls in the dirt.

Start by shearing a flat top the length of the llama's back. Next, taking the shears in one hand, move them in a downward position to remove the coat. Shear a strip the length of the neck from the chin to the front legs about 3 inches wide to help cool the llama. Shearing can take a long time, so it may be necessary for both you and the llama to take a break. Take the llama for a quiet walk and allow it to go to the bathroom so it will not become antsy during the rest of the shearing process.

Collect the sheared fibers in a container and make sure you are working on a clean floor so you can collect any excess fibers and use them for spinning. Do not store the fiber in a plastic bag, as moisture can easily accumulate, ruining the fiber and making it unusable for spinning.

Caring for the Cria

Baby llamas, called cria, require some additional care in their first few days of life. It is important for the cria to receive the colostrum milk from their mothers, but you may need to aid in this process. Approach the mother llama and pull gently on each teat to remove the waxy plugs covering the milk holes. Sometimes, you may need to guide the cria into position under its mother in order for it to start nursing.

Weigh the cria often (at least for the first month) to see that it's gaining weight and growing strong and healthy. A bathroom scale, hanging scale, or larger grain scale can be used for this.

If the cria seems to need extra nourishment, goat or cow milk can be substituted during times when the mother llama cannot produce enough milk for the cria. Feed this additional milk to the cria in small doses, several times a day, from a milking bottle.

Diseases

Llamas are prone to getting worms and should be checked often to make sure they do not have any of these parasites. There is special worming paste that can be mixed in with their food to prevent worms from infecting them. You should also establish a relationship with a good veterinarian who knows about caring for llamas and can determine if there are any other vaccinations necessary in order to keep your llamas healthy. Other diseases and pests that can affect llamas are tuberculosis, tetanus, ticks, mites, and lice.

Using Llama Fibers

Llama fiber is unique from other animal fibers, such as sheep's wool. It does not contain any lanolin (an oil found in sheep's wool); thus, it is hypoallergenic and not as greasy. How often you can shear your llama will depend on the variety of llama, its health, and environmental conditions. Typically, though, every year llamas grow a fleece that is 4 to 6 inches long and that weighs between 3 and 7 pounds. Llama fiber can be used like any other animal fiber or wool, making it the perfect substitute for all of your fabric and spinning needs.

Llama fiber is made up of two parts: the undercoat (which provides warmth for the llama) and the guard hair (which protects the llama from rain and snow). The undercoat is the most desirable part to use due to its soft, downy texture, while the coarser guard hair is usually discarded.

Gathering llama hair is easy. To harvest the fiber, you must shear the llama. However, the steps involved in shearing when you are gathering the fiber are slightly different than when you are simply shearing to keep the llama cooler in the summer months. To shear a llama for fiber collection:

1. Clean the llama by blowing and brushing until the coat is free from dirt and debris.
2. Wash the llama. Be sure to rinse out all of the soap from the hair and let the llama air dry.
3. You can use scissors or commercial clippers to shear the llama. Start at the top of the back, behind the head and neck and work backwards. If using clippers, sheer with long sweeping motions, not short jerky ones. If using scissors, always point them downward. Leave about an inch of wool on the llama for protection against the sun and insect bites. You can sheer just the area around the back and belly (in front of the hind legs and behind the front legs) if your main purpose is to offer the llama relief from the heat. Or you can sheer the entire llama—from just below the head, down to the tail—to get the most wool. Once the shearing is complete, skirt the fleece by removing any little pieces or belly hair from the shorn fleece.

The fiber can be hand-processed or sent to a mill (though sending the fibers to a mill is much more expensive and is not necessary if you have only one or two llamas). Processing the fiber by hand is definitely more cost-effective but you will initially need to invest in some equipment (such as a spinning wheel, drop spindle, or felting needle).

To process the fiber by hand:

1. Pick out any remaining debris and unwanted (coarse) fibers.
2. Card the fiber. This helps to separate the fiber and will make spinning much easier. To card the fiber, put a bit of fiber on one end of the cards (standard wool cards do the trick nicely) and gently brush it until it separates. This will produce a rolag (log) of fiber.
3. Once the fiber is carded, you can use it in a few different ways:

a. Wet felting: To wet felt, lay the fiber out in a design between two pieces of material and soak it in hot, soapy water. Then, agitate the fiber by rubbing or rolling it. This will cause it to stick together. Rinse the fiber in cold water. When it dries, you will have produced a strong piece of felt that can be used in many crafting projects.

b. Needle felting: For this type of manipulation, you will need a felting needle (available at your local arts and crafts or fabric store). Lay out a piece of any material you want over a pillow or Styrofoam piece. Place the fiber on top of the material in any design of your choosing. Push the needle through the fiber and the bottom material and then gently draw it back out. Continue this process until the fiber stays on the material of its own accord. This is a great way to make table runners or hanging cloths using your llama fiber.

c. Spinning: Spinning is a great way to turn your llama fiber into yarn. Spinning can be accomplished by using either a spinning wheel or drop spindle, and a piece of fiber that is either in a batt, rolag, or roving. A spinning wheel, while larger and more expensive, will easily help you to turn the fiber into yarn. A drop spindle is convenient because it is smaller and easier to transport, and if you have time and patience, it will do just as good a job as the spinning wheel. To make yarn, twist two or more pieces of spun wool together.

d. Other uses: Carded wool can also be used to weave, knit, or crochet.

If you become very comfortable using llama fiber to make clothing or other craft items, you may want to try to sell these crafts (or your llama fiber directly) to consumers. Fiber crafts may be particularly successful if sold at local craft markets or even at farmers' markets alongside your garden produce.

Cows

Raising dairy cows is difficult work. It takes time, energy, resources, and dedication. There are many monthly expenses for feeds, medicines, vaccinations, and labor. However, when managed properly, a small dairy farmer can reap huge benefits, like extra cash and the pleasure of having fresh milk available daily.

Breeds

There are thousands of different breeds of cows, but what follows are the three most popular breeds of dairy cows.

The Holstein cow has roots tracing back to European migrant tribes almost two thousand years ago. Today, the breed is widely popular in the United States for their exceptional milk production. They are large animals, typically marked with spots of jet black and pure white.

The Ayrshire breed takes its name from the county of Ayr in Scotland. Throughout the early nineteenth century, Scottish breeders carefully crossbred strains of cattle to develop a cow well suited to the climate of Ayr and with a large flat udder best suited for the production of Scottish butter and cheese. The uneven terrain and the erratic climate of their native land explain the breed's ability to adapt to all types of surroundings and conditions. Ayrshire cows are not only strong and

resilient, but their trim, well-rounded outline, and red and predominantly white color has made them easily recognized as one of the most beautiful of the dairy cattle breeds.

The Jersey breed is one of the oldest breeds, originating from Jersey of the Channel Islands. Jersey cows are known for their ring of fine hair around the nostrils and their milk rich in butterfat. Averaging to a total body weight of around 900 pounds, the Jersey cow produces the most pounds of milk per pound of body weight of all other breeds.

Housing

There are many factors to consider when choosing housing for your cattle, including budget, preference, breed, and circumstance.

Free stall barns provide a clean, dry, comfortable resting area and easy access to food and water. If designed properly, the cows are not restrained and are free to enter, lie down, rise, and leave the barn whenever they desire. They are usually built with concrete walkways and raised stalls with steel dividing bars. The floor of the stalls may be covered with various materials, ideally a sanitary inorganic material such as sand.

A flat barn is another popular alternative, which requires tie-chains or stanchions to keep the cows in their stalls. However, it creates a need for cows to be routinely released into an open area for exercise. It is also very important that the stalls are designed to fit the physical characteristics of the cows. For example, the characteristically shorter Jersey cows should not be housed in a stall designed for much larger Holsteins.

A compost-bedded pack barn, generally known as a compost dairy barn, allows cows to move freely, promising increased cow comfort. Though it requires exhaustive pack and ventilation management, it can notably reduce manure storage costs.

Grooming

Cows with sore feet and legs can often lead to losses from milk production, diminished breeding efficiency, and lameness. Hoof trimming is essential to help prevent these outcomes, though it is often very labor intensive, allowing it to be easily neglected. Hoof trimming should be supervised or taught by a veterinarian until you get the hang of it.

A simple electric clipper will keep your cows well-groomed and clean. Mechanical cow brushes are another option. These brushes can be installed in a free-stall dairy barn, allowing cows to groom themselves using a rotating brush that activates when rubbed against.

Feeding and Watering

In the summer months, cows can receive most of their nutrition from grazing, assuming there is plenty of pasture. You may need to rotate areas of pasture so that the grass has an opportunity to grow back before the cows are let loose in that area again. Grazing pastures should include higher protein grasses, such as alfalfa, clover, or lespedeza. During the winter, cows should be fed hay. Plan to offer the cows 2 to 3 pounds of high quality hay per 100 pounds of body weight per day. This should provide adequate nutrition for the cows to produce 10 quarts of milk per day, during peak production months. To increase production, supplement feed with ground corn, oats, barley, and wheat bran. Proper mixes are available from feed stores. Allowing cows access to a salt block will also help to increase milk production.

Water availability and quantity is crucial to health and productivity. Water intake varies, however it is important that cows are given the opportunity to consume a large amount of clean water several times a day. Generally, cows consume 30 to 50 percent of their daily water intake within an hour of milking. Water quality can also be an issue. Some of the most common water quality problems affecting livestock are high concentrations of minerals and bacterial contamination. Send out 1 to 2 quarts of water from the source to be tested by a laboratory recommended by your veterinarian.

If you intend to run an organic dairy, cows must receive feed that was grown without the use of pesticides, commercial fertilizers, or genetically modified ingredients along with other restrictions.

Breeding

You may want to keep one healthy bull for breeding. Check the bull for STDs, scrotum circumference, and sperm count before breeding season begins. The best cows for breeding have large pelvises and are in general good health. An alternative is to use the artificial insemination (A.I.) method. There are many advantages to A.I., including the prevention of spreading infectious genital diseases, the early detection of infertile bulls, elimination of the danger of handling unruly bulls, and the availability of bulls of high genetic material. The disadvantage is that implementing a thorough breeding program is difficult and requires a large investment of time and resources. In order to successfully execute an A.I. program, you may need a veterinarian's assistance in determining when your cows are in heat. Cows only remain fertile for twelve hours after the onset of heat, and outside factors such as temperature, sore feet, or tie-chain or stanchion housing can drastically hinder heat detection.

Calf Rearing

The baby calf will be born approximately 280 days after insemination. Keep an eye on the cow once labor begins, but try not to disturb the mother. If labor is unusually long (more than a few hours), call a veterinarian to help. It is also crucial that the newborns begin to suckle soon after birth to receive ample colostrum, the mother's first milk, after giving birth. Colostrum is high in fat and protein with antibodies that help strengthen the immune system, though it is not suitable for human consumption. After two or three days, the calf should be separated from the mother and the rest of the herd and placed in a safe, hygienic, and controlled environment to avoid contracting any germs from other animals.

You can teach the calf to drink from a bucket by gently pulling its head toward the pail. A calf should consume about 1 quart of milk for every 20 pounds of body weight. A calf starter can be used to help ensure proper ruminal development. You can find many types of starters on the market, each meeting the nutritional requirements for calves. Begin milking young cows as soon as they are separated from the mother. This will get them used to the process while they are small enough to be more manageable.

Calf vaccination is also very important. You should consult your veterinarian to design a vaccination program that best fits your calves' needs.

Common Diseases

Pinkeye and foot rot are two of the most prevalent conditions affecting all breeds of cattle of all ages year-round. Though both diseases are non-fatal, they should be taken seriously and treated by a qualified veterinarian.

Wooden tongue occurs worldwide, generally appearing in areas where there is a copper deficiency or the cattle graze on land with rough grass or weeds. It affects the tongue, causing it to become hard and swollen so that eating is painful for the animal. Surgical intervention is often required.

Brucellosis or bangs is the most common cause of abortion in cattle. The milk produced by an infected cow can also contain the bacteria, posing a threat to the health of humans. It is advisable to vaccinate your calves, to prevent exorbitant costs in the long run, should your herd contract the disease.

Pigs

Pigs can be farm-raised on a commercial scale for profit, in smaller herds to provide fresh, homegrown meat for your family or to be shown and judged at county fairs or livestock shows. Characterized by their stout bodies, short legs, snouts, hooves, and thick, bristle-coated skin, pigs are omnivorous, garbage-disposing mammals that, on a small farm, can be difficult to turn a profit on but yield great opportunities for fair showmanship and quality food on your dinner table.

Breeds

Pigs of different breeds have different functionalities—some are known for their terminal sire (the ability to produce offspring intended for slaughter rather than for further breeding) and have a greater potential to pass along desirable traits, such as durability, leanness, and quality of meat, while others are known for their reproductive and maternal qualities. The breed you choose to raise will depend on whether you are raising your pigs for show, for profit, or to put food on your family's table.

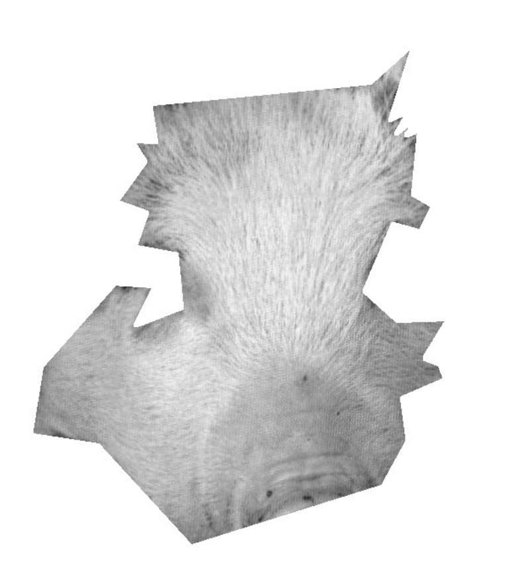

Eight Major U.S. Pig Breeds

1. **Yorkshire**—Originally from England, this large, white breed of hog has a long frame, comparable to the Landrace. They are known for their quality meat and mothering ability and are likely the most widely distributed breed of pig in the world. Farmers will also find that the Yorkshire breed generally adapts well to confinement.
2. **Landrace**—This white-haired hog is a descendent from Denmark and is known for producing large litters, supplying milk, and exhibiting good maternal qualities. The breed is long-bodied and short-legged with a nearly flat arch to its back. Its long, floppy ears are droopy and can cover its eyes.
3. **Chester**—Like the Landrace, this popular white hog is known for its mothering abilities and large litter size. Originating from cross breeding in Pennsylvania, Chester hogs are medium-sized and solid white in color.
4. **Berkshire**—Originally from England, the black and white Berkshire hog has perky ears and a short, dished snout. This medium-sized breed is known for its siring ability and quality meat.
5. **Duroc**—Ranging from solid colors of light gold to dark red, the strong-built, Durocs are known for their rapid growth and ultra efficient feed to meat conversion. This large breed is also hailed for its tasty meat.
6. **Poland China**—Known for often reaching the maximum weight at any age bracket, this black and white breed is of the meaty variety.
7. **Hampshire**—A likely descendent of an Old English breed, the Hampshire is one of America's oldest original breeds. Characterized by a white belt circling the front of their black bodies, this breed is

PIG TERMINOLOGY

pig, hog, or swine	Refers to the species as a whole or any member thereof.
shoat or piglet (or "pig" when species is referred to as "hog")	Any unweaned or immature young pigs.
sucker	A pig between birth and weaning.
runt	An unusually small and weak piglet. Often one per litter.
boar or hog	A male pig of breeding age.
barrow	A male pig castrated before reaching puberty.
stag	A male pig castrated later in life.
gilt	A young female not yet mated (farrowed) or has birthed fewer than two litters.
sow	An active breeding female pig.

TIPS FOR SELECTING BREEDER SOWS

- Look for well-developed udders on a gilt (a minimum of six pairs of teats, properly spaced and functional).
- Do not choose those with inverted teats which do not secrete milk, and do not choose sows that are otherwise unable to produce milk.
- Opt for longer-bodied sows (extra space promotes udder development).

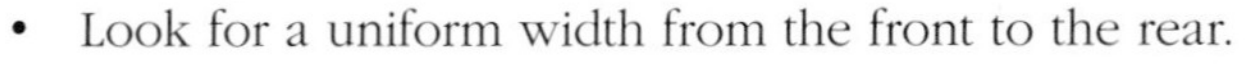

- Look for a uniform width from the front to the rear.
- Check for good development in the ham, loin, and shoulder regions to better assure good breeding.
- Choose the biggest animals within a litter.
- Choose female breeders from litters of eight or more good-sized piglets that have high rates of survivability.
- Choose hardy pigs from herds raised in well-sanitized environments and avoid breeding any pigs with physical abnormalities.

known for its hardiness and high quality meat.

8. **Spot**—Known for producing pigs with high growth rates, this black and white spotted hog gains weight quickly while maintaining a favorable feed efficiency. Part of the Spot's ancestry can be traced back to the Poland China breed.

Housing Pigs

Keeping your pigs happy and healthy and preventing them from wandering off requires two primary structures: a shelter and a sturdy fence. A shelter is necessary to protect your pigs from inclement weather and to provide them with plenty of shade, as their skin is prone to sunburns. Shelters can be relatively simple three-sided, roofed structures with slanted, concrete flooring to allow you to spray away waste with ease. To help keep your pigs comfortable, provide them with enough straw in their shelter and an area to make a wallow—a muddy hole they can lie in to stay cool.

Because pigs will use their snouts to dig and pry their way through barriers, keeping these escape artists fenced in can pose a challenge. "Hog wire," or woven fence wire, at least 40 inches high is commonly used for perimeter fencing. You can line the top and especially the bottom of your fence with a strand of barbed or electric wire to discourage your pigs from tunneling their way through. If you use electric wiring, you may have a difficult time driving your herd through the gate. Covering the gate with non-electric panels using woven wire, metal, or wood can make coaxing your pigs from the pasture an easier task.

Feeding Your Pigs

Pigs are of the omnivorous variety, and there isn't much they won't eat. Swine will consume anything from table and garden scraps to insects and worms to grass, flowers, and trees. Although your pigs won't turn their snouts up at garbage, a cost-effective approach to assuring good health and a steady growth rate for your pigs is to supply farm grains (mixed at home or purchased commercially), such as oats, wheat, barley, soybeans, and corn. Corn and soybean meal are a good source of energy that fits well into a pig's low fiber, high protein diet requirements. For best results, you should include protein supplements and vitamins to farm grain diets.

As pigs grow, their dietary needs change, which is why feeding stages are often classified as starter, grower, and finisher. Your newly weaned piglets make up the starting group, pigs 50–125 lbs. are growing, and those between 125- and the 270-lb. market weight are finishing pigs.

As your pigs grow, they will consume more feed and should transition to a less dense, reduced-protein diet. You should let your pigs self-feed during every stage. In other words, allow them to consume the maximum amount they will take in a single feeding. Letting your pigs self-feed once or twice a day allows them to grow and gain weight quickly.

Another essential part of feeding is to make sure you provide a constant supply of fresh, clean water. Your options range from automatic watering systems to water barrels. Your pigs can actually go longer without feed than they can without water, so it's important to keep them hydrated.

Diseases

You can prevent the most common pig diseases from affecting your herd by asking your veterinarian about the right vaccination program. Common diseases include *E. coli*—a bacteria typically caused by contaminated fecal matter in the living environment that causes piglets to experience diarrhea. You should vaccinate your female pigs for *E. coli* before they begin farrowing.

Another common pig disease is Erysipelas, which is caused by bacteria that pigs secrete through their saliva or waste products. Heart infections or chronic arthritis are possible ailments the bacteria causes in pigs that can lead to death. You should inoculate pregnant females and newly bought feeder pigs to defend against this prevalent disease.

Other diseases to watch out for are Atrophic Rhinitis, characterized by inflammation of a pig's nasal tissues; Leptospirosis, an easily spreadable bacteria-borne disease; and Porcine Parvovirus, an intestinal virus that can spread without showing symptoms. Consult your veterinarian to discuss vaccinating against these and other fast-spreading diseases that may affect your herd.

Butchering

Meat animals may be raised and the meat cured at home for much less than the cost of purchased meat. By raising your own meat, you can ensure that the animals are raised and slaughtered humanely and can avoid the hormones and other health hazards found in most commercially raised meats.

Choosing Animals to Slaughter

Health

In selecting animals for butchering, health should have first consideration. Even though the animal has been properly fed and carries a prime finish, the best quality of meat cannot be obtained if the animal is unhealthy; there is always some danger that disease may be transmitted to the person who eats the meat. The keeping quality of the meat is always impaired by fever or other derangement.

Condition

An animal in medium condition, gaining rapidly in weight, yields the best quality of meat. Do not kill animals that are losing weight. A reasonable amount of fat gives juiciness and flavor to the meat, but large amounts of fat are objectionable.

Figure 1.—Tools for killing and dressing hogs. A, meat saw; B, 14-inch steel; C, Cutting knife; D, hog hook; E, 8-inch sticking knife; F, bell-shaped stick scraper; G, separate parts of stick scraper; H, gambrel.

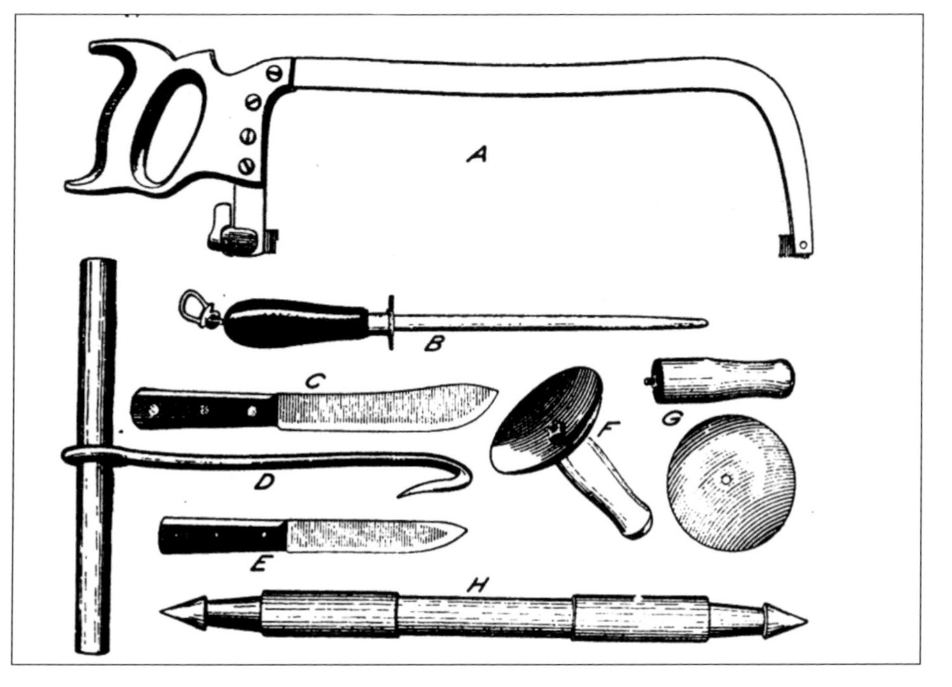

Quality

The breeding of animals plays an important part in producing carcasses of high quality. Selection, long-continued care, and intelligent feeding will produce meat of desirable quality. Smooth, even, and deeply fleshed hogs yield nicely marbled meats.

Age for Killing

The meat from young hogs or cows lacks flavor and is watery, and that from old hogs generally is very tough. However, if older animals are properly fattened before slaughter, the meat will be improved. Hogs or cows may be killed for meat any time after eight weeks of age, but the most profitable age at which to slaughter is between eight and twelve months.

Treatment Before Slaughter

It is easiest to hold cows and pigs entirely without feed for eighteen or twenty-four hours prior to slaughtering, but they should have all the fresh water they will drink. This treatment promotes the elimination of the usual waste products from the system; it also helps to clear the stomach and intestines of their contents, which in turn facilitates the dressing of the carcass and the clean handling and separation of the viscera. Animals should be kept calm as much as possible prior to slaughter.

Killing and Dressing

A .22 caliber gun is best for killing cows and pigs. Male animals should be castrated before slaughtering. If you kill a pig in its pen, immediately afterward throw a noose around its neck and drag it outside to slit the throat from back to front to allow drainage. Cows should be slaughtered outside if they're too large to drag. To sever the main veins and arteries, stick a knife into the throat and cut outward, through the skin. If slaughtering a pig, wash it down at this point.

Use a meat saw to remove the animal's head. Cut slits in the Achilles tendons and insert the gambrel to hoist the animal up to a convenient height, using a pulley or a come-along as needed. The animal should be hanging upside down. For pigs, if you plan to scald (rather than skinning), do so now, then remove the feet at the joint.

Using a sharp, short knife, cut into the slit in the Achilles tendons and down the legs, stopping at the center line. Slice from the center line down to the animal's neck. Begin to remove the skin (unless you used the scalding method), starting where the leg meets the center line and working outward. Leave as much fat intact as possible. Continue slicing around to the front of the leg, working toward the tailbone. Pull the tail sharply to separate the vertebrae.

Cut out the anus in order to remove the intestines. Cut down through the sternum through the belly, and finally between the legs. Be careful not to rupture the bladder. Place a bucket under the animal to catch all the innards. To remove the diaphragm and the heart, you'll need to sever in from the surrounding connective tissue. Finally, remove the windpipe from the neck.

Pork should hang one night in a cold or refrigerated area. Beef can benefit from longer refrigerated aging.

Pigs should be halved before the meat is divided into portions. Cows should be quartered.

Equipment for Dividing the Meat

For cutting up the meat, these old-fashioned tools still get the job done: A straight sticking knife, a cutting knife, a 14-inch steel, a hog hook, a bell-shaped stick scraper, a gambrel, and a meat saw (Figure 1 on previous page). More than one of each of these tools may be necessary if many hogs are to be slaughtered and handled. If you plan to scald a pig rather than skinning it, you can use a large barrel. The barrel should be placed at an angle of about 45 degrees at the end of a table or platform of proper height. The table and barrel should be fastened securely to protect the workmen. A block and tackle will reduce labor. All the tools and appliances should be in readiness before beginning.

Dividing the Meat

Shoulder

Cut off the front foot about 1 inch below the knee. The shoulder cut is made through the third rib at the breastbone and across the fourth. Remove the ribs from the shoulders, also the backbone which is attached. Cut close to the ribs in removing them so as to leave as much meat on the shoulder as possible. These are shoulder or neck ribs and make an excellent dish when fried or baked. If only a small quantity of cured meat is desired, the top of the shoulder may be cut off about one-third the distance from the top and parallel to it. The fat of the shoulder top may be used for lard and the lean meat for steak or roasts. It should be trimmed smoothly. In case the shoulders are very large, divide them crosswise into two parts. This enables the cure mixture to penetrate more easily and therefore lessens the danger of souring. The fat trimmings should be used for lard and the lean trimmings for sausages.

The ham is removed from the middling by cutting just at the rise in the backbone and at a right angle to the shank. The loin and fatback are cut off in one piece, parallel with the back just below the tenderloin muscle on the rear part of the middling. Remove the fat on the top of the loin, but do not cut into the loin meat. The lean meat is excellent for canning or it may be used for chops or roasts and the fatback for lard. The remainder should then be trimmed for middling or bacon. Remove the ribs cutting as close to them as possible. If it is a very large side, it may be cut into two pieces. Trim all sides and edges as smoothly as possible.

Ham

Cut off the foot 1 inch below the hock joint. All rough and hanging pieces of meat should be trimmed from the ham. It should then be trimmed smoothly, exposing as little lean meat as possible, because the curing hardens it. All lean trimmings should be saved for sausage and fat trimmings for lard. The other half of the carcass should be cut up in similar manner.

Loins

Separate the loin from the belly by sawing through the ribs, starting at the point of greatest curvature of the fourth rib. Skin the fat from the loin, leaving the lean muscle barely covered with the fat. The loins can be boned out and used for sausage if a large amount of that is desired, or if the weather will not

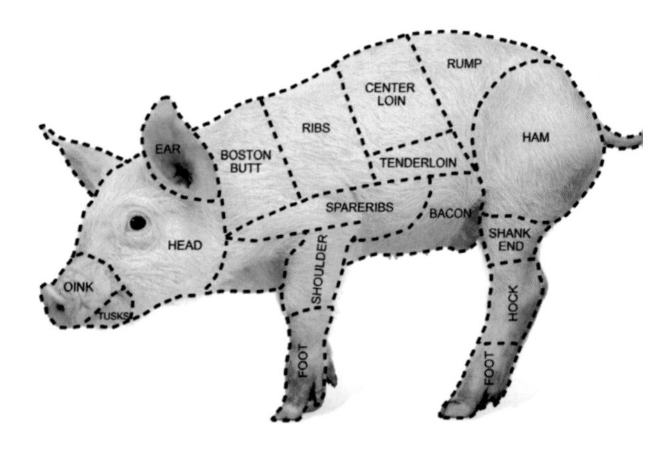

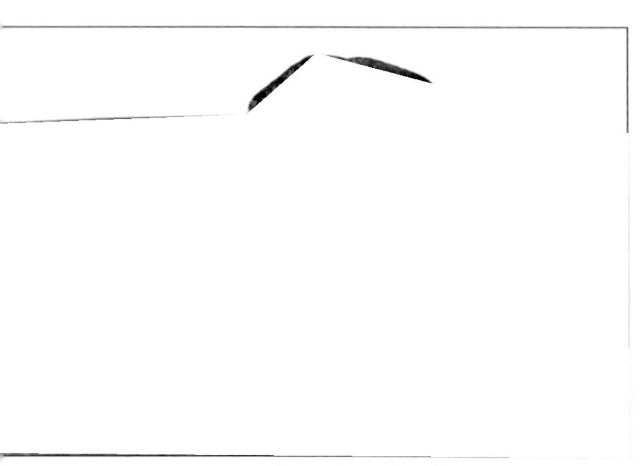

Bacon comes from the pork belly.

permit holding them as fresh meat, they can be given a middle cure as boneless loins. The loin is best adapted for the pork (loin) roast or for pork chops. The latter are cut in such a way as to have the rib end in each alternate piece or chop.

Meat trimmings and fat trimmings

After the carcass has been cut up and the pieces are trimmed and shaped properly for the curing process, there are many pieces of lean meat, fat meat, and fat which can be used for making sausage and lard. The fat should be separated from the lean and used for lard. The meat should be cut into convenient-sized pieces to pass through the grinder.

Rendering lard

The leaf fat makes lard of the best quality. The back strip of the side also makes good lard, as do the trimmings of the ham, shoulder, and neck. Intestinal or gut fat makes an inferior grade and is best rendered by itself. This should be thoroughly washed and left in cold water for several hours before rendering, thus partially eliminating the offensive odor. Leaf fat, back strips, and fat trimmings may be rendered together. If the gut is included, the lard takes on a very offensive odor.

First, remove all skin and lean meat from the fat trimmings. To do this, cut the fat into strips about 1½ inches wide, then place the strip on the table, skin down, and cut the fat from the skin. When a piece of skin large enough to grasp is freed from the fat, take it in the left hand and, with the knife held in the right hand inserted between the fat and skin, pull the skin. If the knife is slanted downward slightly, this will easily remove the fat from the skin. The strips of fat should then be cut into pieces 1 or 1½ inches square, making them about equal in size so that they will dry out evenly.

Pour into the kettle about a quart of water, then fill it nearly full with fat cuttings. The fat will then heat and bring out the grease without burning. Render the lard over a moderate fire. At the beginning, the temperature should be about 160°F, and it should be increased to 240°F. When the cracklings begin to brown, reduce the temperature to 200°F or a little more, but not to exceed 212°F in order to prevent scorching. Frequent stirring is necessary to prevent burning. When the cracklings are thoroughly browned, and light enough to float, the kettle should be removed from the fire. Press the lard from the cracklings. When the lard is removed from the fire, allow it to cool a little. Strain it through a muslin cloth into the containers. To aid cooling, stir it, which also tends to whiten it and make it smooth.

Lard which is to be kept for a considerable amount of time should be placed in air-tight containers and stored in the cellar or other convenient place away from the light, in order to avoid rancidity. Fruit jars make excel-

» A slab of pork lard

SELECTING QUALITY MEAT

As a general rule, the best meat is that which is moderately fat. Lean meat tends to be tough and tasteless. Very fat meat may be good, but is not economical. The butcher should be asked to cut off the superfluous suet before weighing it.

1. *Beef.* The flesh should feel tender, have a fine grain, and a clear red color. The fat should be moderate in quantity, and lie in streaks through the lean. Its color should be white or very light yellow. Ox beef is the best, heifer very good if well fed, cow and bull, decidedly inferior.
2. *Mutton.* The flesh, like that of beef, should be a good red color, perhaps a shade darker. It should be fine-grained, and well-mixed with fat, which ought to be pure white and firm. The mutton of the black-faced breed of sheep is the best, and may be known by the shortness of the shank; the best age is about five years, though it is seldom to be had so old. Whether mutton is superior to either ram or ewe, it may be distinguished by having a prominent lump of fat on the broadest part of the inside of the leg. The flesh of the ram has a very dark color and is of a coarse texture; that of the ewe is pale, and the fat yellow and spongy.
3. *Veal.* Its color should be white, with a tinge of pink; it ought to be rather fat, and feel firm to the touch. The flesh should have a fine delicate texture. The leg-bone should be small; the kidney small and well-covered with fat. The proper age is about two or three months; when killed too young, it is soft, flabby, and dark-colored. The bull-calf makes the best veal, though the cow-calf is preferred for many dishes on account of the udder.
4. *Lamb.* This should be light-colored and fat, and have a delicate appearance. The kidneys should be small and imbedded in fat, the quarters short and thick, and the knuckle stiff. When fresh, the vein in the fore quarter will have a bluish tint. If the vein looks green or yellow, it is a certain sign of staleness, which may also be detected by smelling the kidneys.
5. *Pork.* Both the flesh and the fat must be white, firm, smooth, and dry. When young and fresh, the lean ought to break when pinched with the fingers, and the skin, which should be thin, yield to the nails. The breed having short legs, thick neck, and small head is the best. Six months is the right age for killing, when the leg should not weigh more than 6 or 7 lbs.

lent containers for lard, because they can be completely sealed. Glazed earthenware containers, such as crocks and jars, may be also be used. All containers should be sterilized before filling, and if covers are placed on the crocks or jars, they also should be sterilized before use. Lard stored in air-tight containers away from the light has been found to keep in perfect condition for a number of years.

When removing lard from a container for use, take it off evenly from the surface exposed. Do not dig down into the lard and take out a scoopful, as that leaves a thin coating around the sides of the container, which will become rancid very quickly through the action of the air.

Curing Pork

The first essential in curing pork is to make sure that the carcass is thoroughly cooled, but meat should never be allowed to freeze either before or during the period of curing.

The proper time to begin curing is when the meat is cool and still fresh, or about twenty-four to thirty-six hours after killing. See page 65 for pork curing suggestions.

Sources and Resources

Adams, Joseph H. *Harper's Outdoor Book for Boys*. New York: Harper & Brothers, 1907.

American Heart Association. *How Can I Manage Stress?* americanheart.org/downloadable/heart/1196286112399ManageStress.pdf (accessed June 24, 2009).

American Wind Energy Association. *Wind Energy Fact Sheet*. www.awea.org/pubs/factsheets/HowWindWorks2003.pdf (accessed June 22, 2009).

Andersen, Bruce and Malcolm Wells. *Passive Solar Energy Book*. Build It Solar (2005). www.builditsolar.com/Projects/SolarHomes/PasSolEnergyBk/PSEbook.htm (accessed June 23, 2009).

Anderson, Ruben. "Easy homemade soap." *Treebugger: A Discovery Company*. treehugger.com/files/2005/12/easy_homemade_s.php (accessed June 24, 2009).

Andress, Elizabeth L. and Judy A. Harrison, ed. *So Easy to Preserve*, 5th ed. Athens, GA: The University of Georgia Cooperative Extension, 2006.

Anthony, G.A., F.G. Ashbrook, and Frants P. Lund. U.S. Department of Agriculture. Farmers' Bulletin: "Pork on the Farm: Killing, Curing, and Canning," Issues 1176-1200. Washington: Government Printing Office, 1922.

Autumn Hill Llamas & Fiber. "Llama Fiber Article." *Autumn Hill Llamas & Fiber*. autumnhillllamas.com/llama_fiber_article.htm (accessed June 24, 2009).

Bailey, Henry Turner, ed. *School Arts Book*, vol. 5. Worcester, MA: The Davis Press, 1906.

Beard, D.C. *The American Boy's Handy Book*. NH: David R. Godine, 1983.

Beard, Linda and Adelia Belle Beard. *The Original Girl's Handy Book*. New York: Black Dog & Leventhal Publishers Inc., 2007.

Bell, Mary T. *Food Drying with an Attitude*. New York: Skyhorse Publishing, Inc., 2008.

Bellows, Barbara. "Solar Greenhouse Resources." *ATTRA: National Sustainable Agriculture Information Service* (2009). attra.ncat.org/attra-pub/solar-gh.html (accessed June 24, 2009).

Ben. "My Inexpensive 'Do It Yourself' Geothermal Cooling System." *Trees Full of Money*. www.treesfullofmoney.com/?p=131 (accessed June 29, 2009).

Benton, Frank. U.S. Department of Agriculture. *The Honey Bee: A Manual of Instruction in Apiculture*. Washington: Government Printing Office, 1899.

Brooks, William P. *Agriculture vol. III: Animal Husbandry, including The Breeds of Live Stock, The General Principles of Breeding, Feeding Animals; including Discussion of Ensilage, Dairy Management on the Farm, and Poultry Farming*. Springfield, MA: The Home Correspondence School, 1901.

Bower, Mark. "Building an inexpensive solar heating panel." *Mobile Home Repair* (Aberdeen Home Repair, 2007). www.mobilehomerepair.com/article17c.htm (accessed June 22, 2009).

Boy Scouts of America. *Handbook for Boys*. New York: The Boy Scouts of America, 1916.

"Build a Solar Cooker." *The Solar Cooking Archive*. www.solarcooking.org/plans/default.htm (accessed June 22, 2009).

Burgdorf, David, Thomas Cogger, Glenn Lamberg, Rick Knysz, and Paul Johnson. "Community Garden Guide Season Extension: Hoop Houses." U.S. Department of Agriculture, Natural Resources Conservation Service. http://plant-materials.nrcs.usda.gov/pubs/mipmcarcgghoophouse.pdf (accessed July 1, 2010).

California Integrated Waste Management Board. "Compost—What Is It?" ciwmb.ca.gov/organics/Compost-Mulch/CompostIs.htm (accessed June 24, 2009).

California Integrated Waste Management Board. "Home Composting." ciwmb.ca.gov/Organics/HomeCompost (accessed June 24, 2009).

Call Ducks: Call Duck Association UK. callducks.net (accessed June 24, 2009).

"Candle making." *Lizzie Candles Soap*. lizziecandle.com/index.cfm/fa/ home.page/pageid/12.htm (accessed June 24, 2009).

Comstock, Anna Botsford. *How to Keep Bees; A Handbook for the Use of Beginners*. New York: Doubleday, 1905.

Cook, E.T., ed. *Garden: An Illustrated Weekly Journal of Horticulture in all its Branches*, vol. 64. London: Hudson & Kearns, 1903.

Corie, Laren. "Building a Very Simple Solar Water Heater." *Energy Self Sufficiency Newsletter* (Rebel Wolf Energy Systems, September 2005). www.rebelwolf.com/essn/ESSN-Sep2005.pdf (accessed June 22, 2009).

"Craft instructions: how to make hemp jewelry." *Essortment*. essortment.com/hobbies/makehempjewelr_sjbg.htm (accessed June 24, 2009).

Dahl-Bredine, Kathy. "Windshield Shade Solar Cooker." *Wikia*. solarcooking.wikia.com/wiki/Windshield_shade_solar_funnel_cooker (accessed June 22, 2009).

Dairy Connection Inc. dairyconnection.com (accessed June 24, 2009).

Danlac Canada Inc. danlac.com (accessed June 24, 2009).

Davis, Michael. "How I built an electricity producing Solar Panel." *Welcome to Mike's World*. www.mdpub.com/SolarPanel/index.html (accessed June 22, 2009).

Department of Energy. "Energy Kid's Page." *Energy Information Administration*, November 2007. www.eia.doe.gov/kids/energyfacts/sources/renewable/solar.html (accessed June 26, 2009).

Dickens, Charles, ed. *Household Worlds*, vol. 1. London: Charles Dickens & Evans, 1881.

"DIY Home Solar PV Panels." *GreenTerraFirma*. greenterrafirma.com/home-solar-panels.html (accessed June 23, 2009).

"Do-It-Yourself Wind Turbine Project." *GreenTerraFirma* (2007). greenterrafirma.com/DIY_Wind_Turbine.html (accessed June 23, 2009).

Druchunas, Donna. "Pattern: Fingerless Gloves for Hand Health." *Subversive Knitting*. sheeptoshawl.com (accessed June 24, 2009).

Earle, Alice M. *Home Life in Colonial Days*. New York: Macmillan Company, 1899.

"Easy Cold Process Soap Recipes for Beginners." *TeachSoap.com: Cold Process Soap Recipes*. teachsoap.com/easycpsoap.html (accessed June 24, 2009).

Farmer, Fannie M. *The Boston Cooking-School Cook Book*. Cambridge: University Press, 1896.

Farrington, Edward I. *Practical Rabbit Keeping*. New York: Robert M. McBride and Co., 1919.

Flach, F., ed. *Stress and Its Management*. New York: W.W. Norton & Co., 1989.

"Fun-Panel." *Wikia*. solarcooking.wikia.com/wiki/Fun-Panel (accessed June 22, 2009).

Gegner, Lance. "Llama and Alpaca Farming." *Appropriate Technology Transfer for Rural Areas (ATTRA)*. attra.ncat.org/attra-pub/llamaalpaca.html (accessed June 24, 2009).

Glengarry Cheesemaking and Dairy Supply Ltd. glengarrycheesemaking.on.ca (accessed June 24, 2009).

"Guide to Herbal Remedies." *Natural Health and Longevity Resource Center*. all-natural.com/herbguid.html (accessed June 24, 2009).

Hall, A. Neely and Dorothy Perkins. *Handicraft for Handy Girls: Practical Plans for Work and Play*. Boston: Lothrop, Lee & Shepard Company, 1916.

Hasluck, Paul N. *The Handyman's Book*. London: Cassell and Co., 1903.

Hill, Thomas E. *The Open Door to Independence: Making Money from the Soil*. Chicago: Hill Standard Book Company, 1915.

"Homemade Solar Panel." pyronet.50megs.com/RePower/Homemade%20Solar%20Panels.htm (accessed June 24, 2009).

"Homemade Teat Dip & Udder Wash Recipe." *Fias Co Farm*. fiascofarm.com/goats/teatdip-udderwash.html (accessed June 24, 2009).

"How to Build a Composting Toilet." *eHow, Inc*. www.ehow.com/how_2085439_build-composting-toilet.html (accessed June 28, 2009).

"How to Knit a Hat." *Knitting for Charity: Easy, Fun and Gratifying*. knittingforcharity.org/how_to_knit_a_hat.html (accessed June 24, 2009).

"How to Knit a Scarf for Beginners." *AOK Coral Craft and Gift Bazaar*. aokcorral.com/how2oct2003.htm (accessed June 24, 2009).

"How to Make Hemp Jewelry." *Beadage: All About Beading!* beadage.net/hemp/index.shtml (accessed June 24, 2009).

"How to Make Taper Candles" *How To Make Candles.info* .howtomakecandles.info/cm_article.asp?ID=CANDL0603 (accessed June 24, 2009).

"How to Milk a Goat." *Fias Co Farm*. fiascofarm.com/goats/ how_to_milk_a_goat.htm (accessed June 24, 2009).

"How to Sell Your Crafts on eBay." *Craft Marketer: DIY Home Business Ideas*. craftmarketer.com/sell-your-crafts-on-ebay-article.htm (accessed June 24, 2009).

J.G. "The Fragrance of Potpourri." *Good Housekeeping*, January 1917. New York: Hearst Corp., 1916.

Junket: Making Fine Desserts Since 1874. junketdesserts.com (accessed June 24, 2009).

Kellogg, Scott and Stacy Pettigrew. *Toolbox for Sustainable City Living: A Do-It-Ourselves Guide*. Cambridge, MA: South End Press, 2008.

Kendall, P. and J. Sofos. "Drying Fruits." *Nutrition, Health & Food Safety*. Colorado State University Cooperative Extension: No. 9.309). uga.edu/ nchfp/how/dry/csu_dry_fruits.pdf (accessed June 24, 2009).

Kleen, Emil, and Edward Mussey Hartwell. *Handbook of Massage*. Philadelphia: P. Blakiston Son & Co.,1892.

Kleinheinz, Frank. *Sheep Management: A Handbook for the Shepherd and Student*, 2nd ed. Madison, WI: Cantwell Printing Company, 1912.

Ladies' Work-Table Book, The: Containing Clear and Practical Instructions in Plain and Fancy Needlework, Embroidery, Knitting, Netting and Crochet. Philadelphia: G.B. Zeiber & Co., 1845.

Lambert, A. *My Knitting Book*. London: John Murray, 1843.

Lamon, Harry M. and Rob R. Slocum. *Turkey Raising*. New York: Orange Judd Publishing Company, 1922.

"Learn to Make Beeswax Candles." *MyCraftBook*. mycraftbook.com/Make_Beeswax_Candles.asp (accessed June 24, 2009).

Lindstrom, Carl. *Greywater*. www.greywater.com (accessed June 25, 2009).

Llucky Chucky Llamas. llamafarm.com/welcome.html (accessed June 24, 2009).

Lynch, Charles. *American Red Cross Abridged Text-book on First Aid: General Edition, A Manual of Instruction*. Philadelphia: P. Blakiston's Son & Co., 1910.

"Make Your Own Paper." *Environmental Education for Kids!* dnr.wi.gov/org/caer/ce/eek/cool/paper.htm (accessed June 24, 2009).

"Marketing your homemade crafts." *Essortment*. essortment.com/all/craftsmarketing_mfm.htm (accessed June 24, 2009).

McGee-Cooper, Ann. *You Don't Have to Go Home From Work Exhausted!: The energy engineering approach*. Dallas, Texas: Bowen & Rogers, 1990.

Moore, Donna. "Shear Beauty." *International Lama Registry*. lamaregistry.com/ilreport/2005May/shear_beauty_may.html (accessed June 24, 2009).

Moorlands Cheesemakers: Suppliers of Farm and Household Dairy Equipment. cheesemaking.co.uk (accessed June 24, 2009).

Mountain, Johnny. "Raising Rabbits." http://www.thefarm.org/charities/i4at/lib2/rabbits.htm (accessed July 1, 2010).

Morais, Joan. "Beeswax Candles." *Natural Skin and Body Care Products*. naturalskinandbodycare.com/2008/12/beeswax-candles.html (accessed xxxx xx, xxxx).

Murphy, Karen. "How to make beeswax candles." *SuperEco*. supereco.com/how-to/how-to-make-beeswax-candles (accessed June 24, 2009).

Natural Skin and Body Care Products. naturalskinandbodycare.com (accessed June 24, 2009).

N., Beth. "How to Make Taper Candles." *Associated Content*. associatedcontent.com/article/360786/how_to_make_taper_candles.html?cat=24 (accessed June 24, 2009).

National Ag Safety Database. "Basic First Aid: Script." *Agsafe*. nasdonline.org/docs/d000101-d000200/d000105/d000105.html (accessed June 24, 2009).

National Center for Complementary and Alternative Medicine. "Herbal Medicine." *MedlinePlus: Trusted Health Information for You*. nlm.nih.gov/medlineplus/herbalmedicine.html (accessed June 24, 2009).

National Center for Complementary and Alternative Medicine. *Herbs at a Glance*. nccam.nih.gov/health/herbsataglance.htm (accessed June 24, 2009).

National Center for Complementary and Alternative Medicine. *Massage Therapy: An Introduction*. nccam.nih.gov/health/massage/#1 (accessed June 24, 2009).

National Center for Home Food Preservation. "Drying: Herbs." uga.edu/nchfp/how/dry/herbs.html (accessed June 24, 2009).

National Center for Home Food Preservation. "General Freezing Information." uga.edu/nchfp/how/freeze/dont_freeze_foods.html (accessed June 24, 2009).

National Center for Home Food Preservation. "USDA Publications: USDA Complete Guide to Home Canning, 2006." uga.edu/nchfp/publications/publications_usda.html (accessed June 24, 2009).

National Institutes of Health: Office of Dietary Supplements. "Botanical Dietary Supplements: Background Information." *Office of Dietary Supplements*. ods.

od.nih.gov/factsheets/BotanicalBackground.asp (accessed June 24, 2009).

National Renewable Energy Laboratory. "Wind Energy Basics." *Learning About Renewable Energy*. www.nrel.gov/learning/re_wind.html (accessed June 24, 2009).

New England Cheesemaking Supply Company. cheesemaking.com (accessed June 24, 2009).

Nissen, Hartvig. *Practical Massage in Twenty Lessons*. Philadelphia: F.A. Davis Company, 1905.

Nucho, A. O. *Stress Management: The Quest for Zest*. Illinois: Charles C. Thomas, 1988.

Nummer, Brian A. "Fermenting Yogurt at Home." National Center for Home Food Preservation: uga.edu/nchfp/publications/nchfp/factsheets/yogurt.html (accessed June 24,2009).

Ostrom, Kurre Wilhelm. *Massage and the Original Swedish Movements: Their Application to Various Diseases of the Body*. 6th ed. Philadelphia: P. Blakiston's Son & Co., 1905.

Ponder, T. *How to Avoid Burnout*. Mountainview, CA: Pacific Press Publishing Association, 1983.

Reyhle, Nicole. "Selling Your Homemade Goods." *Retail Minded*. retailminded.com/blog/2009/01/selling-your-homemade-goods (accessed June 24, 2009).

Retail Minded. retailminded.com/blog (accessed June 24, 2009).

Sanford, Frank G. *The Art Crafts for Beginners*. New York: The Century Co., 1906.

Sell, Randy. "Llama" *Alternative Agriculture* Series, no. 12. ag.ndsu.edu/pubs/alt-ag/llama.htm (accessed June 24, 2009).

Sheep to Shawl. sheeptoshawl.com (accessed June 24, 2009).

Sherlock, Chelsa C. *Care and Management of Rabbits*. Philadelphia: David McKay Co., 1920.

Singleton, Esther. *The Shakespeare Garden*. New York: The Century Co., 1922.

Smith, Kimberly. "Where to Sell Your Homemade Crafts Offline." *Associated Content*. associatedcontent.com/article/1678550/where_to_sell_your_homemade_crafts.html (accessed June 24, 2009).

"Soap making–General Instructions." *Walton Feed, Inc*. waltonfeed.com/old/old/soap/soap.html (accessed June 24, 2009).

"Soy candle making." *Soya–Information about Soy and Soya Products*. soya.be/soy-candle-making.php (accessed June 24, 2009).

Swenson, Allan A. *Foods Jesus Ate and How to Grow Them*. New York: Skyhorse Publishing, Inc., 2008.

Szykitka, Walter. *The Big Book of Self-Reliant Living: Advice and Information on Just About Everything You Need to Know to Live on Planet Earth*, 2nd ed. Guilford, CT: The Lyons Press, 2004.

Taylor, George Herbert. *Massage: Principles and Practice of Remedial Treatment by Imparted Motion*. New York: John B. Alden, 1887.

Thompson, Nita Norphlet and Sue McKinney-Cull. "Soothing Those Jangled Nerves: Stress Management." *ARCH Factsheet*, no. 41. archrespite.org/archfs41.htm (accessed June 24, 2009).

U.S. Department of Agriculture: Food Safety and Inspection Service. *Fact Sheets: Egg Products Preparation*. fsis.usda.gov/Factsheets/Focus_On_Shell_Eggs/index.asp (accessed June 24, 2009).

U.S. Department of Agriculture: Food Safety and Inspection Service. *Fact Sheets: Poultry Preparation*. fsis.usda.gov/Fact_Sheets/Chicken_Food_Safety_Focus/index.asp (accessed June 24, 2009).

U.S. Department of Agriculture: Natural Resources Conservation Service. "Backyard Conservation: Composting." nrcs.usda.gov/feature/backyard/compost.html (accessed June 24, 2009).

U.S. Department of Agriculture: Natural Resources Conservation Service. "Backyard Conservation: Nutrient Management." nrcs.usda.gov/feature/backyard/nutmgt.html (accessed June 24, 2009).

U.S. Department of Agriculture: Natural Resources Conservation Service. "Composting in the Yard." nrcs.usda.gov/feature/backyard/compyrd.html (accessed June 24, 2009).

U.S. Department of Agriculture: Natural Resources Conservation Service. "Home and Garden Tips: Composting." nrcs.usda.gov/feature/ highlights/homegarden/compost.html (accessed June 24, 2009).

U.S. Department of Agriculture: Natural Resources Conservation Service. "Home and Garden Tips: Lawn and Garden Care." nrcs.usda.gov/feature/highlights/homegarden/lawn.html (accessed June 24, 2009).

U.S. Department of Agriculture: National Agricultural Library. "Organic Production." afsic.nal.usda.gov/nal_display/index.php?info_center= 2&tax_level=1&tax_subject=296 (accessed June 24, 2009).

U.S. Department of Energy. "Active Solar Heating." *Energy Efficiency and Renewable Energy: Energy Savers*. www.energysavers.gov/your_home/space_heating_cooling/index.cfm/mytopic=12490 (accessed June 26, 2009).

U.S. Department of Energy. "Benefits of Geothermal Heat Pump Systems." *Energy Efficiency and Renewable Energy: Energy Savers*. www.energysavers.gov/your_home/space_heating_cooling/index.cfm/mytopic=12660 (accessed June 25, 2009).

U.S. Department of Energy. "Energy Efficiency and Renewable Energy." *Wind and Hydropower Technologies Program*. www1.eere.energy.gov/windandhydro/ (accessed June 24, 2009).

U.S. Department of Energy. "Energy Technologies." *Efficiency and Renewable Energy: Solar Energy Technologies Program*. www1.eere.energy.gov/solar/want_pv.html (accessed June 26, 2009).

U.S. Department of Energy. "Geothermal Heat Pumps." *Energy Efficiency and Renewable Energy: Energy Savers*. www.energysavers.gov/your_home/space_heating_cooling/index.cfm/mytopic=12650 (accessed June 26, 2009).

U.S. Department of Energy. "Heat Pump Water Heaters." *Energy Efficiency and Renewable Energy: Energy Savers*. www.energysavers.gov/your_home/water_heating/index.cfm/mytopic=12840 (accessed June 26, 2009).

U.S. Department of Energy. "Hydropower Basics." *Energy Efficiency and Renewable Energy: Wind and Hydropower Technologies Program*. www1.eere.energy.gov/windandhydro/hydro_basics.html (accessed June 26, 2009).

U.S. Department of Energy: National Renewable Energy Laboratory. "Direct Use of Geothermal Energy." *Office of Geothermal Technologies*. www1.eere.energy.gov/geothermal/pdfs/directuse.pdf (accessed June 26, 2009).

U.S. Department of Energy: National Renewable Energy Laboratory. "Wind Energy Myths." *Wind Powering American Fact Sheet Series*. www.nrel.gov/docs/fy05osti/37657.pdf (accessed June 26, 2009).

U.S. Department of Energy. "Renewable Energy." *Energy Efficiency and Renewable Energy: Energy Savers*. www.energysavers.gov/renewable_energy/solar/index.cfm/mytopic=50011 (accessed June 26, 2009).

U.S. Department of Energy. "Selecting and Installing a Geothermal Heat Pump System." *Energy Efficiency and Renewable Energy: Energy Savers*. www.energysavers.gov/your_home/space_heating_cooling/index.cfm/mytopic=12670 (accessed June 25, 2009).

U.S. Department of Energy. "Technologies." *Energy Efficiency and Renewable Energy: Geothermal Technologies Program*. www1.eere.energy.gov/geothermal/faqs.html (accessed June 25, 2009).

U.S. Department of Energy. "Solar." *Energy Sources*. www.energy.gov/energysources/solar.htm (accessed June 26, 2009).

U.S. Department of Energy. "Toilets and Urinals." *Greening Federal Facilities*, second edition. www.eere.energy.gov/femp/pdfs/29267-6.2.pdf (accessed June 29, 2009).

U.S. Department of Energy. "Your Home." *Energy Efficiency and Renewable Energy: Energy Savers*. www.energysavers.gov/your_home/space_heating_cooling/index.cfm/mytopic=12300 (accessed June 26, 2009).

U.S. Environmental Protection Agency. "Composting Toilets." *Water Efficiency Technology Fact Sheet*. www.epa.gov/owm/mtb/comp.pdf (accessed June 29, 2009).

U.S. House of Representatives. United States Department of Agriculture. *Report of the Commissioner of Patents for the Year 1831: Agriculture*. 37th congress, 2nd sess., 1861.

University of Maryland. *National Goat Handbook*. uwex.edu/ces/cty/richland/ag/documents/national_goat_handbook.pdf (accessed June 24, 2009).

"Where to sell crafts? Consider these often overlooked alternative markets..." *Craft Marketer: DIY Home Business Ideas*. craftmarketer.com/where_to_sell_crafts.htm (accessed June 24, 2009).

Whipple, J. R. "Solar Heater." *J. R. Whipple & Associates*. www.jrwhipple.com/sr/solheater.html (accessed June 23, 2009).

Wickell, Janet. *Quilting*. Teach Yourself Books. Chicago: NTC Publishing Group, 2000.

Williams, Archibald. *Things Worth Making*. New York: Thomas Nelson and Sons, Ltd., 1920.

"Wind Energy Basics." *Wind Energy Development Programmatic EIS*, windeis.anl.gov/guide/basics/index.cfm (accessed June 25, 2009).

Wolok, Rina. "How to Build a Composting Toilet." *Greeniacs*, June 15, 2009. greeniacs.com/GreeniacsGuides/How-to-Build-a-Composting-Toilet.html (accessed June 29, 2009).

Woods, Tom. "Homemade Solar Panels." *Forcefield* (2003). www.fieldlines.com/story/2005/1/5/51211/79555 (accessed June 24, 2009).

Woolman, Mary S. and Ellen B. McGowan. *Textiles: A Handbook for the Student and the Consumer*. New York: The Macmillan Company, 1921.

Worcester Polytechnic Institute. "A Passive Solar Space Heater for Home Use." *Solar Components Corporation*. www.solar-components.com/SOLARKAL.HTM#doityourself (accessed June 22, 2009).

Young Ladies' Journal, The: Complete Guide to the Work-Table. London: E. Harrison, 1885.

Index

A
aphids, 42, 44
apiculture. See beekeeping
asparagus, 47

B
baking tips, 53, 58
basketweaving
  birch bark basket, 194
bath salts and scrubs, 191
beans, 47
beekeeping
  bee pastures, 225
  beeswax, 226
  hives
    building, 224
    making, 227
    situating, 224
  honey extraction, 225
  reasons for, 223
  stings, avoiding, 224
  swarming, 24
  tips, 226
  winter, preparing bees for, 6, 225
berry ink, 175
birds, butterflies, and bees
  plant species for, 33
blackberries, 27
bleeding heart plants, 8
boomerangs, 83
bread, making, 56
breads, 53
broccoli, 48
Brussels sprouts, 48
butchering, 249
butter, 72
making, 72

C

cabbage, 48
cake wedding, 172
candles, 184
- essential oils for mosquito repellants, 188
- floating, 187
- gourd votives, 187
- jarred soy candles, 187
- natural dyes for, 187
- rolled beeswax candle, 184
- taper candles, 185

canning, 82
- altitude, 87
- apple butter, 99
- apple juice, 97
- applesauce, 100
- apricots, 102
- beans
  - baked, 125
  - green, 126
  - shelled or dried, 124
- beets, 127
- benefits of, 83
- berries, 102
- berry syrup, 104
- bread-and-butter pickles, 143
- canners, 91
  - boiling-water canners, 92
  - pressure canners, 92
- carrots, 128
- controlling headspace, 89
- corn
  - cream style, 129
  - whole kernel, 130
- dill pickles, 139, 144
- ensuring high-quality, 87
- food acidity and processing methods, 86
- fruit, 96
- fruit purées, 105
- fruit spreads, 122
- glossary, 84
- grape juice, 105
- hot packing, advantages, 88
- how canning preserves food, 83
- identifying and handling spoiled canned food, 95
- jams, jellies, and other fruit spreads, 113
  - blueberry-spice jam, 121
  - grape-plum jelly, 121
  - jam without added pectin, 118
  - jellies without added pectin, 114
  - lemon curd, 116
  - peach-pineapple spread, 122
  - pear-apple jam, 119
  - refrigerated apple spread, 123
  - refrigerated grape spread, 123
  - remaking soft jellies, 124
  - strawberry-rhubarb jelly, 120
- jars and lids, 89
  - cleaning, 90
  - cooling, 95
  - lid selection, preparation, and use, 90
  - seals, 95
  - sterilization of empty jars, 90
- maintaining color and flavor, 88
- marinated peppers, 142
- meat stock (broth), 136
- mixed vegetables, 130
- peaches, 106
- pears, 107
- peas, 132
- piccalilli, 142
- pickle relish, 145
- pickled horseradish sauce, 141
- pickled three-bean salad, 140
- pie fillings, 109
  - apple pie, 111
  - blueberry pie, 111
  - cherry pie, 112
  - mincemeat pie, 113
- potatoes, sweet, 132
- practices, proper, 86
- processing time, 91
- pumpkin, 134
- rhubarb, 107
- sauerkraut, 140
- soups, 134
- special diets, 96
- spiced apple rings, 100
- storing canned food, 95
- succotash, 134
- sweet pickles, quick, 145
- sweet pickles, reduced-sodium sliced, 146
- syrups, 97
- tomatoes
  - chile salsa (hot tomato-pepper sauce), 153
  - crushed, 148
  - juice, 147
  - ketchup, 153
  - sauce, 149
  - spaghetti sauce without meat, 152
  - whole or halved, packed in water, 150
- what not to do, 87

cantaloupe, 48
carrots, 48
cauliflower, 48
cheese, 75
- cheddar, 79
- cheese press,78
- from goat's milk, 231
- making, 75
- mozzarella, 78
- preparation, 76
- queso blanco, 77
- ricotta cheese, 77
- yogurt cheese, 76

chickens
- breed selection, 209
- chicks
  - hatching, 210
- eggs from, 211
- feeding, 209
  - free-range, 208
- housing, 208
  - coops, 209
- meat from, 209
- slaughtering, 219

collards, 48
companion planting, 6
composting
- barrels used for, 16
- making your own, 16
- materials to use, 16
- preparation of, 15
- problems, 17
- types of
  - cold or slow, 17
  - vermicomposting, 17
- uses for, 18

containers
- flowers and, 35
- herbs and, 34
- preserving plants in, 35
- things to consider, 34, 35
- vegetables and, 33

corn, heirloom, 48
cornhusk dolls, 192
cows
- breeding, 243
- breeds, 241
- butchering, 250
- calf rearing, 243
- diseases, 244
- feeding and watering, 243
- grooming, 242
- housing, 242

crackers, 80
crafts, 165
coffee cake, 54
cucumbers, 48
Cutworms, 42, 44

D

dandelion, 2
dolls, cornhusk, 192
dried plants, 167
drying foods, 155
- food dehydrators, 155
  - woodstove dehydrator, 159

fruit drying procedures, 156
- fruit leathers, 157
- herbs, 159
- jerky, 160
- procedures, 156

- pumpkin leather, 159
- tomato leather, 159
- vegetable leathers, 158

ducks
- breeds, 212
- diseases, 215
- ducklings
  - caring for, 215
  - hatching, 214
- feeding and watering, 214
- housing, 213
- slaughtering, 219

E

eggplant, 48
eggs
- blown, 167
- leaf- or flower-stenciled, 168
- natural dyes for Easter, 168

elderberry, 26, 102

F

farmers' markets, 223
feather pens, 175
flour, 57
flower pomander, hanging, 172
flowerpots, mosaic, 171
flowers
- growing in containers, 35
- preserving, 36

flowers, preserving, 179
- natural wax flowers, 179
- pressed flowers and leaves, 179

freezing foods, 162
- foods that do not freeze well, 163
- fruits, 163
- meat, 164
- vegetables, 162

fruit bushes and trees
- blueberries, 27
- brambles and bush fruits, 27
- currants and gooseberries, 28
- fruit growing chart, 26
- fruit trees, 28
- grapes, 29
- strawberries, 25

G

gardening tools, 6
gardening, choosing a site for, 5
- elevation, 6
- irrigation, 6
- proximity, 5
- rain, 6
- soil quality, 6, 9
- sunlight and, 2, 5
- water availability, 6

germination temperatures, 21
goats
- breeds, 228
  - angora, 232
- cheese, 231
- diseases, 230
- feeding,
- milk, 228

grains, 229
- growing and threshing, 30
- milling, 58

H

hams
- country-style ham, 68
- regular-cut ham, 68
- virginia ham
  - aging, 69
  - carving, 71
  - cooking, 71
  - curing, 67
  - deboning,
  - pests, 69
  - preparing, 70

harvesting, 47
home gardening, 2
Homeschool Hints
- sprouting seeds, 21

houses,
- chicken, 208
- henhouse, 208

I

ice cream
- making, 80

ink quill pens, berry, 175
insects. See pests

J

Japanese beetles, 41
Junior Homesteader
- composting, 17
- gardens, 8
- how soap works, 191
- plant playhouses, 5

K

kale, 48
knots, 181
- bowline, 181
- clove hitch, 181
- halter, 181
- sheepshank knot, 181
- slip knot, 182
- square/reef knot, 182
- timber hitch, 182
- two half hitches, 182

L

labeling rows, 23
lampshades, 194
lettuce, 48
lima beans, 48
llamas, 237
- cria, 239
- diseases, 239
- feeding, 238
- fibers from, 239
- housing, 238
- shearing, 238
- toenail trimming, 238

locally grown food, 83

M

maple sugaring, 61
- tapping of, 61
- making of, 62

maple syrup, 61
marigold, 45
mint, 7
mustard, 49

N

nettle, 192

O

okra, 49
onions, 49
open sheds, 234
organic, 11

P

paper, handcrafted, 203
pasteurizing milk, 74
peas, 49
peppers, 49
periwinkle, 8
pests and disease management, 41
- disease identification, 44
  - powdery mildew leaf disease, 46
- identifying the problem, 41
- insects and mites, 41
  - aphids, 42, 44
  - beneficial insects, 44
  - cutworms, 42, 44
  - Japanese beetles, 41
  - worm-eaten apples, 43
- Integrated pest management (IPM), 42
- natural repellants, 44, 45
- practices, 45

pies
- apple pie filling, 111
- blueberry pie filling, 111
- cherry pie filling, 112
- festive mincemeat pie filling, 113

pigs
- breeds, 245
- butchering, 250
- diseases, 248
- feeding, 248

housing, 247
terminology, 246
pine cone bird feeder, 180
plant nutrients
management of, 10
testing, 10
plants
basic needs of, 2
cold, 3
heat, 3
length of day, 2
light, 2
soil pH, 3
temperature, 3
water, 3
for decorative use, 167
glossary of, 4
shade-loving, 7
bleeding heart plants, 8
flowering plants, 8
vegetable plants, 8
potatoes, 49
potpourri, 176
rose potpourri, 178
sachet potpourri, 178
"you choose" potpourri, 177
pottery
basic vase, 199
basics of, 198
bowls, 202
candlesticks, 201
decorating, 202
embellishments, 200
firing, 202
glazing, 202
jars, 201
sawdust kiln, 202
vases, 202
wheel-working, 199
poultry, slaughtering, 219
pumpkins, 49

Q

quick breads, 52
basic quick bread recipe, 53
cinnamon bread, 53
cranberry coffee cake, 54
date muffins, 55
date-orange bread, 54
one-hour brown bread, 54
pineapple nut bread, 55
quilting, 196

R

rabbits
breeding, 221
breeds, 220
feeding, 221
health concerns, 222
housing, 221
radishes, 49
raised beds, 38
how to make, 39
things to consider, 40
raspberries, 26, 27
raw milk, 74
rope, knots, 181
rugs, rag, 206
rutabagas, 49

S

sachets, 179
sausage making, 64
beef sausage, 66
bratwurst, cooked, 67
important considerations in, 64
polish sausage (kielbasa), 66
pork sausage, 65, 252
venison or game sausage, 66
venison or game sausage, 66
scrapbooking, 204
seedlings, 19
sheep
breeds, 233
butchering, 253
diseases, 236
feeding, 234
housing, 234
milk, 236
shearing, 235
wool
carding, 235
spinning, 236
soap making, 188
cold-pressed soap, 188
natural dyes for, 192
soap oils for, 190
soil
enriching, 12
with compost, 12
with fertilizers, 13
nutrients
macronutrients, 10
management of, 10
micronutrients, 10
testing, 11
pH, 3, 10
quality of, 9
types of
clay, 22
loam, 22
sandy, 22
watering,
spinach, 50
springtime wreath, 166
sprouting seeds, 20
squash, 50
sweet potatoes, 50
swiss chard, 50
strawberry, 25

T

temperature zone map, 3
terrariums, 169
tomatoes, 50
turkeys
breeds, 216
chicks ("poults")
hatching, 217
raising, 218
diseases, 218
feeding, 217
housing, 217
slaughtering, 219
turnips, 50

V

vegetables
making your own, 23
things to consider, 24
venison or game sausage, 66
violets, 8, 174

W

walnut, 113
water, 3
weddings, 171
bouquets, 171
cake, 172
centerpieces, 173
edible flowers, 173
hanging flower pomander, 172
napkins, 174
programs for the ceremony, 174
sachets, 173
wheat, 55
worms, 16, 17
wreaths, 166

Y

yeast, 56
yeast bread, 55
biscuits, 56
gluten-free bread, 60
making your own, 56
multigrain bread, 58
oatmeal bread, 59
raised buns (brioche), 59
tips, 58
wheat and, 55
yeast and, 56
yogurt, 73
making, 73
types, 75